ABERDEEN UNIVERSITY STUDIES SERIES
Number 155

The Fishing Industries of Scotland, 1790–1914

The Fishing Industries of Scotland, 1790–1914

A Study in Regional Adaptation

MALCOLM GRAY

Published for the UNIVERSITY OF ABERDEEN
by
OXFORD UNIVERSITY PRESS
1978

Oxford University Press, Walton Street, Oxford OX2 6DP

OXFORD LONDON GLASGOW
NEW YORK TORONTO MELBOURNE WELLINGTON
IBADAN NAIROBI DAR ES SALAAM LUSAKA CAPE TOWN
KUALA LUMPUR SINGAPORE JAKARTA HONG KONG TOKYO
DELHI BOMBAY CALCUTTA MADRAS KARACHI

British Library Cataloguing in Publication Data
Gray, Malcolm
    The fishing industries of Scotland, 1790–1914.
    1. Fisheries – Scotland – History
    I. Title
    338.3'72'709411   SH259   78–40244

ISBN 0–19–714105–6

*Printed in Great Britain by
Cox & Wyman Ltd,
London, Fakenham and Reading*

Preface

THIS BOOK owes much to my colleagues of the Department of Economic History at Aberdeen University. Perhaps without knowing it, they have helped in many ways—not only in the discussions of the formal seminar but also in the exchange of ideas, of criticisms and of perceptions that form the stuff of a departmental collaboration; just as important they have encouraged me to believe that I have something to say on this subject. James Coull has helped me greatly, through his writings and in discussion, out of his wide knowledge of the problems of fishing in many countries. Donald Withrington, miraculously in view of the calls on his time, has given the text a thorough revision that was more than editorial in its scope; I have not agreed with all his suggestions on style but undoubtedly he has managed to eradicate many obscurities, ambiguities and awkward expressions. Pat Smith and Lynne Cockburn, in typing and sometimes re-typing the text, have succeeded in reducing an often chaotic manuscript to order and neatness and I am deeply grateful to them. The cartographers of the Geography Department similarly turned my sketchy indications into maps and charts of professional finish. Like almost everyone who has researched to any depth in Scottish history I find myself much in debt to the staff of the Scottish Record Office for their invariable helpfulness; we in Scotland are indeed fortunate to have at our service such patience and sympathetic help. Finally, Susan Gray has relieved me of much tedious but basic and essential work on the records; more important perhaps, she has invariably supported me by patience and by encouragement through the difficult phases of the work.

M. GRAY

Aberdeen, 1977

Contents

LIST OF MAPS AND FIGURES ix

ABBREVIATIONS x

INTRODUCTION 1

I. A Traditional Fishing Economy: the East Coast
in the 1790s 9

II. The Rise of Herring Fishing: Caithness, 1790–1815 27

III. The Widening Sphere of Herring Fishing, 1815–1835 39

IV. Herring Fishing Dominant: the Driving Forces,
1835–1884 58

V. The East Coast Fisherman, 1835–1884 80

VI. Crofting and Fishing: the West Coast, 1790–1884 101

VII. Crofting and Fishing: the Northern Isles,
1800–1880 124

VIII. Herring at its Peak: the Older Communities,
1884–1914 146

IX. The New Community: Trawling, 1880–1914 166

X. The Crofting Communities in the New Era,
1884–1914 181

Conclusion 210

BIBLIOGRAPHY 219

INDEX 225

List of Maps

Fishing settlements, North East Area, 1855   11
Fishing settlements, Forth Area, 1855   12
Fishing settlements, Caithness and Sutherland, 1855   28
North West Area   102
Clyde Area   118
Shetland   125
Orkney   125

List of Figures

1   Production and Export of Cured Herring   40
2   Wick: Price per barrel of cured herring (Crown Brand
    Fulls)   59
3   Stettin: Imports of cured herring from all sources and from
    Scotland   62
4   Fraserburgh: Rate paid per cran of uncured herring   72
5   Wick: Average catch per boat at the Caithness summer
    herring fishing   82
6   Fraserburgh: Annual increase in number of herring boats   83
7   Fraserburgh: Ratio of catch to value per boat (boat and
    gear)   84
8   Fraserburgh: Rate paid per cran of uncured herring   93
9   Fraserburgh: Average catch per boat at summer herring
    fishing   93
10 Fraserburgh: Average gross earnings per boat (summer
    herring fishing)   94
11 Fraserburgh: Estimated income per (owning) fisherman
    (summer herring fishing)   94
12 Fraserburgh: Value of boats and gear per owning fisherman   96
13 East Coast: Annual increase in total value of boats and gear   96
14 Fraserburgh: Index of equipment per head (real terms)   97
15 Fraserburgh: Price paid for uncured herring (per cran)   150
16 Price per barrel of cured herring (Crown Brand Fulls)   150

Abbreviations

BFS   British Fisheries Society
FBR   Fishery Board Records
FB Rep. Fishery Board Annual Reports
NSA   New Statistical Account
OSA   Old Statistical Account (Statistical Account of Scotland)
SRO   Scottish Record Office

Introduction

SCOTLAND IS almost surrounded, except along her short border with England, by seas which are among the most prolific fishing areas of the world. Around her coasts for a considerable distance offshore stretch waters of no more than moderate depth—less than 100 fathoms—and in them thrive the species adapted to cool water conditions of less than 200 fathoms; particularly important in such conditions are cod and haddock. Furthermore, the edge of the continental shelf is near enough for the catching, by boats fitted out for only a short sojourn at sea, of fish which may seek deeper water, such as the ling. Furthermore, the movement of the surface water created by the meeting of currents from Atlantic and the Baltic sustains a plankton supply which brings in the pelagic herring in great numbers. Thus, Scotland's fishing history is almost entirely to be written in terms of three species or groups of fish —the herring, the haddock and the 'great fish' such as cod and ling.

Particularly fortunate has been Scotland's position in relation to the prevailing movements of the herring. For long it has been known that shoals are likely to appear annually at places and times that give the appearance of a great circulating movement around the country—from the winter concentration on the west coast, past Shetland in spring and early summer, to the east coast grounds in late summer, and ultimately off the English coast in autumn. In fact, the appearance of a single mass of fish on the move is misleading. The herring which appear on different parts of the coast at the successive seasons of the year are of different sub-species, each with its particular system of migration. But whatever may be the explanation, the intermittent but broadly predictable appearances have given annual opportunities for herring fishing successively over different parts of the Scottish coastline almost throughout the year, while fishermen all round the country have had the chance, not always seized, of a nearby herring fishing at some time of the year. This opportunity derives mainly from the propensity of herring to gather in shoals for spawning. The main stock perhaps is that which arrives for spawning from July to September along various sections of the east coast. This concentration is part of a sweeping movement from over-wintering grounds near Norway through feeding grounds to the south. Herring of a different stock, deriving it appears from Atlantic sources, gather round the west coast islands and mainland generally in the winter months, but this general area also supports some spring spawners. The Shetland Islands lie in an area of meeting between Atlantic and Norwegian stocks and can provide fishing opportunities almost continuously from May to September.

This pattern is to be understood as a definition of possibilities within broad limits, and has been subject to much change both year by year and over longer periods; the annual size of the stock and the exact position of spawning have tended to frequent and often irregular fluctuation and, in particular, the variations in the proximity of the shoals to the shores have meant perplexity for fishermen. These changes have been controlled largely by the movement and accumulation of the surface plankton which are, in their turn, responsive to the mingling and relative strength of currents of a different temperature from the Atlantic and the Baltic. Geographically remote causes may affect this hydrographic movement and it is small wonder that the variations in herring fishing have been not only sharp but also utterly unpredictable and even, in retrospect, inexplicable.

The cod is a fish of different, and ultimately more dependable, habit. Generally keeping to depths of less than 200 fathoms it is wide-ranging and adaptable to different water temperatures. For most of its life it does not appear to follow definite migration patterns, although at spawning, in which fish of four years old and more participate, the species gathers in shoals mainly in the deeper water. It is then—in the period of the year between February and April—that it can be caught most plentifully; the spawning grounds tend to be in remote and difficult areas but even in the era of small boats some Scottish fishermen were able to reach them. There follows for the young fish a period of feeding in nursery areas which include the littoral of Scotland. Apart from the spawning and nursery periods the cod will range widely, regardless of water temperatures, in search of food which it finds mostly, but not entirely, on the sea-bed; as a predator, however, it tends to follow the herring for prey, and the herring's movements, as we have seen, are determined by its food supplies on the water's surface. Almost all over the continental shelf cod is found as a ground-feeding fish, and few parts of the Scottish coastline do not provide the opportunity of catching the fish by methods which reach down to the sea-bed. The ling—a similar great fish—haunts rather deeper water and can best be caught in a westerly direction at the edge of the shelf; for some west coast fishing areas, and more particularly for Shetland, it offers better prey than the cod.

Of peculiar importance in Scottish fishing habit and tradition has been the haddock. Like the cod it gathers for spawning in shoals mainly in the deeper parts of the North Sea. Again, there are grounds directly to the east and north both of the mainland and of the Outer Hebrides that can be reached by the shore-based Scottish fishermen. Thereafter, the young fish spread to nursery grounds mainly in the shallower waters in the western part of the North Sea, thus giving the opportunity for fishing close inshore. Finally, the mature fish returns to the deeper water. On the whole, the haddock has a lesser tolerance of cold water than the cod and the greatest densities are found in the southern parts of the North Sea. Altogether, the haddock, like the cod, spreading along the coasts of Scotland, provide stocks which can be widely fished both in inshore and deeper waters.

The opportunities for developing a successful fishing industry are determined not only by the nearby environmental conditions of water and sea-bed but also by facilities for landing, launching and sheltering boats; by proximity of the landing areas to markets; by natural features which allow processing. Scotland stands out as a country with, for its size, an exceptionally long coastline, but the coastal outline and overall shape of the country have made for complexity and variation. Different parts of the coastline have very different potential qualities as fishing bases.

Because of the shape and orientation of the country, the coastline is largely comprised in the two conventional divisions of east and west coasts. The distinction is more than one of geographical location, for the two main divisions are different in many of their features. The east coast is one of long sweeps breasting the sea, the inlets being broad and open firths. The only exception to this is the Firth of Cromarty which, almost closed at the mouth, widens on the landward side into an expansive and sheltered anchorage. Thus, the dominant feeling almost everywhere along the coast must be one of exposure, and any shelter granted to boats must be an artificial one. The coastal outline is, in fact, the edge of an agricultural plain which stretches almost uninterrupted from the northern to the southern extremities. This varies greatly in extent, from the forty or more miles of breadth found in the north-east to a strip a mere mile in width in Sutherland. The land is intensively cultivated, has supported a big agricultural population, and is dotted with small towns which are the commercial and communications centres of an economically active area. Three of the large cities of Scotland lie on this coast, and in Fife and southwards there has long been a fairly dense mining and industrial population. North and south communication by land has been developed so that each point on the coast is connected to arteries that lead into the great population centres.

The west coast, on the other hand, is fretted with deep and narrow inlets, and in the south the long sea lochs, mainly running north and south, are given added protection by the long arm of the Kintyre peninsula. The lochs of the whole coast not only provide sheltered water and safe anchorage, but also form miniscule and separate fishing areas. The islands which lie offshore in the two chains of the Inner and Outer Hebrides repeat these coastal features and provide farther distinct fishing areas although, particularly on the extreme western or Atlantic seaboard, long, bare sweeps are to be found. The position of the island mass in relation to the mainland has also created a much larger version of the pattern seen in miniature in the sea loch—the partially enclosed stretch of water which is virtually surrounded by shores on which fishermen can establish their bases. The fishing grounds, then, are of three types—within the lochs, in the Minch, and in the open Atlantic to the west on the belt bordered by the edge of the continental shelf.

Not all the features of the west coast favour the fisherman. The lochs and inlets cut deep into a mountain mass and the population is squeezed on to narrow coastal ledges, not at all continuous, around the outline of the coast. Thus, north of Kintyre local markets are limited and connection with more distant markets is difficult. The main landward communications run east and west and the best contacts are made with adjacent centres on the east, but many sectors find even this communication difficult and the route to the great population centres is long and tortuous. The great exception to this is the Firth of Clyde region east of Kintyre, from every part of which it is only a short trip by water to reach the massive industrial concentrations of south-west Scotland. Further south still, in Ayrshire, is an environment more like that of the east coast. The main markets of the islands—which in the relation of sea to land reproduce most of the features of north-west mainland—were in the Glasgow region and were reached by sea.

In the northern isles there are detached fragments which, in coastal pattern, conform more to the west coast than to the east. The deep indentation of mainland blocks coupled with the siting of smaller islands placed irregularly across some of the paths of entry to the mainland coastline create, once again, extensive areas of sheltered water which can be used both as bases for the boats and also for the fishing itself. The land is very different on the two groups of islands. The Orkneys, flat and mainly cultivable, have provided the basis for large and specialized agricultural population, but in Shetland the rough and peaty land-surface confines agricultural working within congested coastal areas.

Within this geographical framework, the development by 1800 of the fishing industry in Scotland had been determined partly by broad strokes of national policy. Eighteenth-century policy developed in large measure as a reaction to the success of the Dutch fishing industry. Indeed, the shadow of the Dutch—with a herring fishing that spread along the coasts of Europe and a commercial control reaching towards central and eastern Europe—lay across the aspirations of other nations to successful fishing. Thus in 1750, after the previously dominant fishing of Fife died away, a new coherent attempt at encouragement was made, with the adoption by the government of the bounty system. By this scheme, which was to last in its broad principles for eighty years, bounties were promised for vessels fitting out for the herring fishing, on a scale related to the size of the vessel. There was much regulation about the manner, place and time of fishing but no payment was made relating to the size of the catch. In one way the measure was a success, in that every year a fleet, which ultimately comprised over 200 vessels, was engaged in fishing for and curing herring: on the whole, activity and output tended to increase as time went on. But in other more important respects it was a failure. For the fleet never rose above its dependence on the bounty; it was only the extra payments that allowed any profit to be made or, indeed, led to the continuance of any activity at all. It proved an effort of shallow roots. The merchants who took the financial risks were men for whom herring fishing was but one of many interests; the vessels were designed for more than one purpose and were turned into trading ships for a large part of the year; and the labour force was created by attracting, for a season, men from different districts and occupations. When profits failed the whole enterprise was threatened with extinction, leaving no solid interest or community. But, at the end of the century, public money still brought the annual creation of a fishing fleet, artificial and rootless as it might be.

The public interest in fishing was also expressed in the foundation in 1786 of the British Fisheries Society, which was in fact a joint-stock company promoted by men of the landlord and merchant classes with an over-riding interest in social development, particularly of the western Highlands. Their method was to create villages—or to set the conditions for the growth of villages—in which fishermen would congregate. This got much nearer than did the bounty legislation to the men for whom fishing would be a major and continuing part of livelihood but, until they turned to the east coast in the nineteenth century, only one of their three west-coast settlements, Ullapool, had any success as a fishing village. There did seem to be emerging a new kind of community, holding land and also devoted to a rather hazardous herring fishing; but the promise was killed by the retreat of the herrings in the early nineteenth century, and in any case this was never more than a miniscule effort in comparison either with the bounty fleet or with the thousands who carried on fishing as a series of tiny private enterprises without government aid or public interest.

For there were, in fact, hardier and more durable forms of fishing lying beyond the narrow gaze of the legislators. Nearly every day of every year thousands of men worked their small boats out of simple harbours, lochs or unprotected creeks, all round the coasts. Such men were handicapped rather than helped by the legislator. Prohibition was laid on the sale to the 'busses'—the vessels fitted out under the terms of the bounty—of herring caught in smaller boats. And dispersed fishing effort was gravely injured by the operation of the salt laws. Fish curing required the use of foreign salt on which outrageous rates of duty were levied. True, fish-curers were entitled to import such salt duty-free but the practical use of this privilege was so complicated that only determined curers, men with money and time, were likely to take it up. Bonds had to be given for salt imported for use in the fishery; customs houses—which might be far distant from the seat of the fishery—had to be visited; and the salt had to be kept under lock and key.

Thus, until nearly the end of the century, the ordinary fishermen were hampered rather than helped by legislation. But in 1786 came the signs of a change. An act in that year promised bounties, at the rate of 2s. per barrel, on herring caught from small boats and cured to conform with the regulations. This, added to 2s. 4d. due on each barrel of herring exported, gave a substantial subsidy to all forms of herring fishing and the effect of general encouragement was strengthened ten years later when the salt laws were made less stringent, salt being allowed to be kept without the use of a locked store-room. Thus fishing moved into the century which was to see its great expansion to that pre-eminence—even over the Dutch—of which Scotsmen had long dreamed.

A further factor in the development of fishing was the nature of the social order in which the fishermen were enmeshed. The differences of social organization that were evident from place to place tended to fall into a broad regional pattern, with lines of division following those of geography. There was a typical west coast form of society, an east coast one and yet another in the northern isles.

In the west were to be found small and fairly isolated communities, generally with scanty resources of arable land. Land hunger was intense and the whole social organization was shaped to provide inevitably small portions of land to each family. Fishing, then, took on the rôle of providing as many families as possible with some extra income and subsistence. Smallholdings, widely distributed, left most men with some free time to pursue fishing but few with the opportunity or desire to spend their whole time thereat. Moreover, general poverty and prior calls upon available capital led to many difficulties in achieving full efficiency, while the generally powerful landlord class failed to provide any consistent support to fishing.

In the east it was different. The communities of fisherman and farmer tended to separate and the fisherman, with no other livelihood, had a different point of view and a different grasp of opportunities from the farmer-fisherman of the west. With his very existence depending on regular and reasonable returns from his fishing, the drive to find the best system of fishing was correspondingly more intense; the opportunities of continuing activity through every season could be exhausted; even remote fishing grounds could be explored; and incomes could be squeezed to provide some savings for the predominant object of investment.

The northern isles, particularly Shetland, conformed to the west coast system of social organization with wide distribution of smallholdings which provided less than full subsistence. But an important difference was the single-mindedness of the landlords in supporting and profiting from the fishing.

Variations of social organization as between east, west and north, together with the inevitable differentiations of geography, were to be continuing influences on development. Every region had its particular combination of influences and its own direction and pace of development. Our discussion, then, must be presented in terms of continuing distinct systems to be found in the various regions of the country.

CHAPTER I

A Traditional Fishing Economy: the East Coast in the 1790s

I

THE EIGHTEENTH century, particularly in its later years, heard much lamentation about the state of the Scottish fisheries and about the failure to create an industry which would rival that of Holland, a country with less easy access to the main fishing grounds. Yet the discussion paid singularly little attention to the men who, alone in the country, had been successful in sustaining a long-term living from fishing—the inhabitants of the communities of the east coast, in many of which fishing had been the main source of livelihood since at least the sixteenth century. It is true that by the last quarter of the eighteenth century not all of these long-established communities were in flourishing condition. A decline in their traditional fishing afflicted the fishermen of the East Neuk of Fife and in the last decades of the century a fairly general fall in the supply of haddocks brought difficulty to some of the old-established communities. On the other hand, the signs along the Moray Firth were of expansion since the third decade of the century; new villages were founded at various points and the fishing population in this general area was almost certainly greater at the end of the century than it had been at the beginning.[1]

Eastern Scotland had little shelter for boats on its wide bare sweeps of coastline, yet in 1800 was studded with fishing communities over most of its length. These were spaced along the coast at unequal intervals, with a marked clustering on some stretches.[2] The area containing the greatest fishing population was the North East and it was said that from the mouth of the Spey to Montrose—a distance of 140 miles—there were, in the last quarter of the eighteenth century, some seventy distinct communities. Even in this area there were considerable empty stretches —usually where a sandy shore made it difficult for boats to work unless

---

[1] James R. Coull, 'Fisheries in the North-East of Scotland before 1800', *Scottish Studies,* 13 (1969), 20.
[2] *OSA*, III, 282; IV, 537; V, 38; IX, 338, 446; XII, 46.

an artificial harbour was built—but there were also two stretches of particularly dense settlement, the western section of the Banffshire coast and the extreme north-east corner.[1] Another short stretch with a big fishing population was the East Neuk of Fife, where the fishing communities were interwoven in the burghal background. Elsewhere the communities were more widely spaced but there was no very extended section of coast which lacked its fishermen.[2]

The term 'village' is perhaps too pretentious to describe either the scale or the social composition of most of the settlements where the fishers lived. Some had as few as one or two boats and in the 1790s out of forty-eight for which definite figures are available, only four had ten or more (excluding small yawls), the largest being St. Monance and Buckie with fourteen boats each.[3] Very nearly half the villages had between five and seven boats and the village of Portessie, in Banffshire, may be taken as typical; with five large boats and seven yawls, it contained 178 inhabitants occupying forty-five houses.[4] The cottages of the fishing community, usually less than fifty in number, conformed in type to the poorer agricultural buildings of the time, being usually set in a closely-huddled group either, on the seashore or in the folds and on the summits of the cliffs and then reached by path from the shores on which the boats lay. By the end of the eighteenth century a more definite concept of planning was beginning to be applied to the newer villages, with regular streets and squares, but irregularly spaced cottages, gable-end to the sea, were still typical.[5]

Small as these groups might be, they were also notably close-knit, being visibly separate from the populations of farm and burgh which surrounded them on every landward side, Some were compact sections of larger towns, but even then the fishers lived apart in well-defined areas; the 'fishertown' or the 'seatown' was a familiar part of many small Scottish burghs of the east coast and within them lived, in chosen seclusion, the people of the fishing community. The burgh of Fraserburgh had about 1,000 inhabitants; Broadsea, to the north of the town, contained the fishing community which had seven boats and 200 people.[6] Fishermen were often regarded by their neighbours outside the fishertown almost as an alien people.

[1] Coull, 'Fisheries in the North-East of Scotland', 21.
[2] BFS, Vol. 3/146, 'Report to the Board of Trustees on the white fishing trade', 1787.
[3] OSA, IX, 338; XIII, 400.
[4] OSA, XIII, 401.
[5] Coull, 'Fisheries in the North-East of Scotland', 22.
[6] OSA, VI, 9.

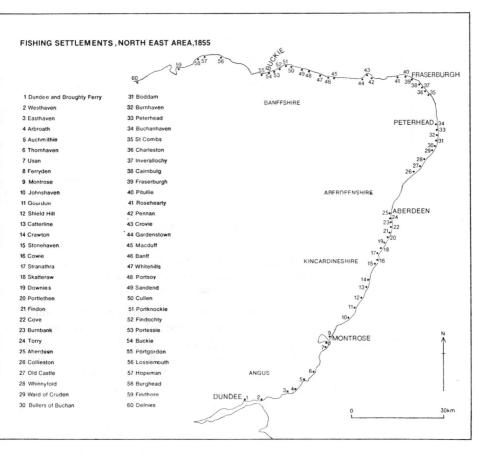

FISHING SETTLEMENTS, NORTH EAST AREA, 1855

1 Dundee and Broughty Ferry	31 Boddam
2 Westhaven	32 Burnhaven
3 Easthaven	33 Peterhead
4 Arbroath	34 Buchanhaven
5 Auchmithie	35 St Combs
6 Thornhaven	36 Charleston
7 Usan	37 Inverallochy
8 Ferryden	38 Cairnbulg
9 Montrose	39 Fraserburgh
10 Johnshaven	40 Pitullie
11 Gourdon	41 Rosehearty
12 Shield Hill	42 Pennan
13 Catterline	43 Crovie
14 Crawton	44 Gardenstown
15 Stonehaven	45 Macduff
16 Cowie	46 Banff
17 Stranathra	47 Whitehills
18 Skatteraw	48 Portsoy
19 Downies	49 Sandend
20 Portlethen	50 Cullen
21 Findon	51 Portknockie
22 Cove	52 Findochty
23 Burnbank	53 Portessie
24 Torry	54 Buckie
25 Aberdeen	55 Portgordon
26 Collieston	56 Lossiemouth
27 Old Castle	57 Hopeman
28 Whinnyfold	58 Burghead
29 Ward of Cruden	59 Findhorn
30 Bullers of Buchan	60 Delnies

Source: Parliamentary Papers, 1856, XVIII, Boat Account.

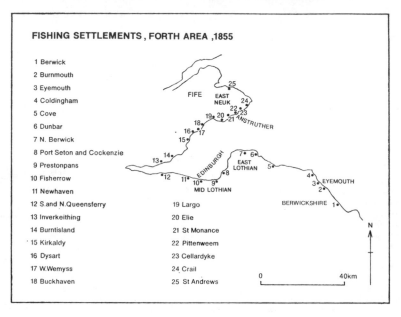

FISHING SETTLEMENTS, FORTH AREA ,1855

1 Berwick
2 Burnmouth
3 Eyemouth
4 Coldingham
5 Cove
6 Dunbar
7 N. Berwick
8 Port Seton and Cockenzie
9 Prestonpans
10 Fisherrow
11 Newhaven
12 S.and N.Queensferry
13 Inverkeithing
14 Burntisland
15 Kirkaldy
16 Dysart
17 W.Wemyss
18 Buckhaven

19 Largo
20 Elie
21 St Monance
22 Pittenweem
23 Cellardyke
24 Crail
25 St Andrews

Source: Parliamentary Papers, 1856, XVIII, Boat Account.

'The fishers should be placed in a quarter of the Town by themselves which they like, and others following other employments do not like to be near them as the smell when boiling and making their oil is most unpleasant.'[1]

Much more common, however, was the small fishing community standing on its own, with farmers, cottagers and labourers as the nearest neighbours. Up to then, fishing villages might be not far off and several stretches of eastern coastline had strings of villages lying within sight of each other. Banffshire, for example, had six villages on ten miles of coast from Cullen to Portgordon; Aberdeenshire had four in the parish of Cruden; around Fraserburgh there were six on the north-eastern crook of coast.[2] But however closely they might lie to each other they remained a series of distinct if tiny units both with clear physical boundaries and with a clear community identity.

The fishers were, indeed, socially as well as geographically distinct from people of other occupations. The fishermen tended to be in small groups, which were recruited solely from fishing families and in which few who had been born to such a family would take up any other occu-

[1] BFS, 306/1 'Observations regarding fishers and fishing villages', 1787.
[2] OSA, V, 99, 433; VI, 9; XII, 145; XIII, 400–2; XVI, 633.

pation and the general absorption of members of fishing communities in the one main occupation sprang from the traditional continuities of a family life in which almost invariably son followed father to a place in an established crew. New recruits to fishing were generally absorbed one by one in existing crews of older men; and normally the initial link was that of blood relationship, with, most commonly of all, son joining father in the boat. Each established crew had its place for junior members—the 'boy' who would perform special operations and draw a minor share from the proceeds. 'They go to sea as boys at 14 years of age, become men at 18 and marry soon after.'[1] 'Becoming a man' meant taking a full share in a boat and, with marriage and a share in the boat, the youngster had the means of life-long livelihood with the fullest possible standing within the fishing community. Almost invariably, then, the son would follow the father to sea; and nearly all the recruits to fishing were drawn in due succession from existing fishing families.

The exclusiveness of fishing communities was maintained not only in the recruitment of boats' crews from among the families of the previous generation of owners but also in marriage customs. A fisherman would seldom marry other than a girl from a fishing family.[2] Indeed, the lot of the fisherman's wife was so notoriously hard that few were likely to accept the position who had not been brought up to fishing.

The fisher wives lead a most laborious life. They assist in dragging the boats on the beach, and in launching them. They sometimes, in frosty weather, and at unseasonable hours, carry their husbands on board, to keep them dry. They receive the fish from the boats, carry them, fresh or after salting, to their customers, and to market at the distance, sometimes, of many miles, through bad roads and in a stormy season. . . . It is the province of the women to bait the lines; collect furze, heath or the gleanings of the mosses, which, in surprising quantity, they carry home in their creels for fuel, to make the scanty stock of peats and turfs prepared in summer, last till the returning season.[3]

This was the life of a fisherman's wife in Banffshire, in a northerly area where most of the fish had to be cured, where the boats had little direct access to market and where evidently the household depended on fuel gathered by personal labour. Such forms of toil might be absent in the more southerly region where much of the fish was sold fresh in a market to which the boats might sail directly. Yet the basic toil of

[1] *OSA*, XIII, 423–4.
[2] BFS, Vol. 3/80, 'Extracts from letters', 1787; *OSA*, XIII, 423–4; XV, 630–1; XVI, 18–19, 517.
[3] *OSA*, XIII, 424.

baiting lines and of handling boats was always there, and in every part of the eastern seaboard the fisherman had the reputation of marrying only his own kind. In Buckhaven, for example, where there were also weavers and miners in the village, the fishermen still took wives from among their own kind.[1]

Fishing was pursued on the east coast as a permanent and full-time occupation absorbing the whole of each succeeding generation to the exclusion of other pursuits. 'From Duncansby Head in Caithness the fishermen all along from that South to Berwick are hereditary and regular bred fishers from their infancy.'[2] It may not always have been so. It is possible, even probable, that fishing communities had originally been created by a slow parturition within agricultural settlements. They mostly came under the same landlords as the agricultural tenantry; common names occur between the fishing and agricultural groups; even in 1800 they are found occupying residual portions of land.[3] Yet it is clear that even north of the Tay, where their whole background was more completely agricultural than it was in the south, by the end of the eighteenth century the fishers were engaged in their occupation full-time and had but slight interest in the working of land. The Buckie people we are told in 1787 'care but little about land or gardens'.[4] In general where fishers held land it was in tiny plots used only for potato planting. Sometimes they had sloughed off the working of the land while retaining the formal rights of occupation. Thus, in one area by the Moray Firth the fishers held potato ground to which the farmers 'carry out (dung), give the land two ploughings and harrowings and keep the land clean'.[5] Consequently, the fishers 'have sufficiency for their families and to fatten two pigs, the only four-footed animals they keep'.[6] In fact, any interest in the land involving any considerable use of time would have represented a conflict with the main interest of their lives for in the annual calendar the seasonal fishings were so closely interlocked, one with another, that the true fishermen had no time for any other occupation. Sometimes the women-folk would turn to agricultural work, but even then they were still much involved in the tasks of fishing. There were other slight breaks in the monolithic devotion to fishing even within the fishing community. In Johnshaven, for instance, the men

[1] OSA, XV, 517.
[2] BFS, Vol. 3/80, 'Extracts from letters', 1787.
[3] Coull, 'Fisheries in the North-East of Scotland', 23–5.
[4] BFS, 306/1, 'Observations regarding fishers and fishing villages', 1787.
[5] John Shirreff, General View of the Agriculture of the Shetland Islands (Edinburgh, 1814), 79-80.
[6] Shirreff, Shetland, 80.

might go to sea for a period; and in Arbroath fishermen also acted as pilots.[1] But these are small exceptions to the general rule of specialization in fishing mitigated only by the tilling, from place to place, of vestigial fragments of land.

II

The fishing engaged in by these specialist communities was universally a small-boat fishing, to grounds which could be reached at a few hours' journeying from the home village, the normal fishing base. The use of small boats was, in practice, imposed by the lack of harbour accommodation or natural anchorage which meant that in many places only boats small enough to be hauled over the beaches could be used. All the way from the Cromarty Firth to the English border, with the exception of the inner reaches of the Firths of Tay and of Forth—which were in any case too far from fishing grounds to be widely used—every mile of the coastline lay exposed to the lash of the winds and the beat of the waves of a notoriously unpredictable sea. Nor were any of the fishing communities big enough to bear the cost of constructing an artificial harbour through the revenue which it generated. Some fishermen could take advantage of a commercial harbour through which there passed a more general trade. But only in Fife were the harbours numerous enough to serve any but a minority of fishermen. Most had to work from open creeks, tiny channels through which the boats would pick a way to a small area of land having a relatively smooth surface and gentle slope across which they could be drawn beyond reach of tide. This operation of landing and launching conditioned the manner of fishing.

Usually the east coast fishermen alternated two different types of fishing through the successive seasons of the year. For most of the year they would seek to capture haddock and other small fish, mainly on inshore grounds, reached by a brief journey from a shore base which would also be their home. Each trip lasted only a few hours and daily the boats would move to the grounds, lay and haul the lines and return to shore, all within the hours of daylight. The means of capture was the 'small-line'. In such a line several strings would be tied end-to-end to form a single main line of between 300 and 1,500 fathoms. To the main line were attached 'snoods', or short lines of thinner calibre on each of which was fixed a hook. The snoods were set three or four feet apart and altogether a line making up a boat's operational gear would have

---

[1] James Headrick, *General View of the Agriculture of the County of Angus, or Forfarshire* (Edinburgh, 1813), 97.

anything between 500 and 3,000 hooks.[1] The number of strings, the length of the line and the number of hooks varied according to the custom of the particular district or even village and sometimes a different length of line would be used in summer and in winter.

There were detailed differences, too, in the method of operation but the general principles were everywhere the same. Lines when in use were cast out so that one end was anchored and marked by a buoy; then gradually, by the skilful movement of the boat, the remainder was paid out to lie on the bottom of the sea-bed, one end being held on the boat. After a period the line was hauled to the surface, being taken in as the boat was moved back to the original point of anchorage. Fish, of course, had to be removed while the boat was directed with pace and direction under control and the line was either dropped in loose coils on the bottom of the boat or was more carefully fitted into the original basket or box-like wooden container.

The laborious task of baiting, usually with mussels, was normally performed ashore—as we have seen—by the fishermen's wives. Always there would be at least two sets of lines so that one could be baited while the other was in use, but sometimes three sets were used so that each line had its sequence of baiting, use and drying.[2] Baiting and drying were performed in or near the cottage and it might be the additional task of the wives to carry the lines to and from the boat, as well as to help with the launching and hauling above high-water mark.

Haddock fishing, being dependent on the preparation of the lines ashore, was limited to one cycle of shooting and hauling of lines on each trip and much of the fishing was conducted in inshore waters. For such short trips small yawls of no more than 16-foot keel might be used but in some cases larger boats of 20-foot keel or more were worked, either because they served both for the haddock and for more distant types of fishing or because the haddock grounds themselves were at some distance.[3]

Much of the haddock was sold fresh and in Fife and the districts south, with the advantage of the Edinburgh markets, none at all would be cured in any way.[4] But farther north, while there was generally some local sale of fresh haddocks, much of the fish was cured. By 1800 the

---

[1] James Arbuthnot, *An Historical Account of Peterhead from the earliest period to the present time* (Aberdeen, 1815), 38–40; *OSA*, VII, 204–5; XIII, 403.

[2] J. Y. Mather, 'Aspects of the Linguistic Geography of Scotland: III. Fishing Communities of the East Coast (Part 1)' *Scottish Studies*, xiii, 1969, 5–8.

[3] BFS, Vol. 3/146, 'Report to the Board of Trustees on the white fishing trade', 1787; Shirreff, *Shetland Islands*, 80–81.

[4] *OSA*, I, 124; III, 116; IV, 370; IX, 338; XVI, 516.

smoking of haddocks was already an established craft in which some villages had a particular reputation. Notably, the tiny village of Findon with only two boats, was already giving its name to a type of cure in which the haddock, having been split so that they presented a flat surface, were smoked over peat fires within the household 'lum'.[1]

Curing by smoking was the custom elsewhere along the Buchan and Moray Firth coasts, but sometimes in the summer the fish would be subjected to the 'spelding' form of cure, in which light salting was combined with smoking and drying in the sun.[2] But whatever the local peculiarities of the method of curing it was everywhere a household occupation, and much of the cured (like the fresh) fish was sold direct to the consumer by the fishermen's wives who might tramp to the inland districts to find their customers.[3] There was also a substantial export of cured haddocks out of the region.[4]

Fishing for small fish such as haddocks, whiting and codling, mostly in inshore waters was, in nearly all east coast communities, the occupation of the greater part of the year and particularly of autumn and winter. Some villages, such as those of the inner Moray Firth, had no other fishing; but the east coast fishermen would most frequently also spend part of the year in the pursuit of the great fish such as cod and ling,[5] which were taken in more distant waters and were generally most profitably fished in the late spring. Thus at some time in the first half of the year the crews would prepare for this fishing, which required its own apparatus and for which very often a larger type of boat—of up to 25-foot keel—would be kept.[6] From Fife they would make for the Mar Bank seawards of the Isle of May, from the Buchan coast it was the 'Forties' ground, the Moray Firth crews would strike in a northeasterly direction, and from Eyemouth they went down beyond the Holy Island.[7] Altogether these distant fishings would last for between three and five months.

The means of capture was the great-line, an apparatus of the same

[1] BFS, Vol. 3/148, 'Report on the white fishing trade', 1787; Headrick, *Angus*, 98.
[2] BFS, Vol. 3/148, 'Report on the white fishing trade', 1787.
[3] *OSA*, V, 277–8; VIII, 206; James Anderson, *General View of the Agriculture and Rural Economy of the County of Aberdeen* (Edinburgh, 1794), 35.
[4] BFS, Vol. 3/148, 'Report on the white fishing trade,' 1787.
[5] BFS, Vol. 3/145, 'Report on the white fishing trade', 1787.
[6] David Souter, *General View of the Agriculture of the County of Banff* (Edinburgh, 1812), 80; Shirreff, *Shetland*, 81; BFS, Vol. 3/146, 'Report on the white fishing trade', 1787.
[7] Arbuthnot, *Historical Account of Peterhead*, 38; *OSA*, XIII, 402; XVI, 549.

fundamental principle as the small-line but of heavier type and carrying fewer hooks.[1] The hooks would be baited on the trip with herring or haddock that would be caught as a preliminary to the main fishing. Because lines could be shot and hauled several times on the one trip and because the distance to the grounds made it wasteful of time to return frequently, the boats would go to the grounds for two or three days at a time.

Cod and ling were mostly cured by drying in the wind and sun. This was a lengthy process but, with landings accruing in steady and small quantities, the whole operation could be completed by the fishermen's families, each dealing with its own particular share of the catch. Thus in Rathven in Banffshire, 'cod, ling and tusk are salted in pits on the beach as they are caught and dried on the rocks for sale'.[2] The drying on the rocks was in fact a long and skilful process in which the fish had to be laid out, turned to expose the skin and flesh for just the right period and periodically built into 'steeples' in which the individual fish were frequently interchanged in position so that each was submitted to the same pressures over the period. In the autumn the whole year's output of dried fish would be taken to one of the market centres of the south of Scotland.[3] Some cod, however, were sold uncured, from the boat, to merchants who in the late eighteenth century were beginning to arrive in the fishing districts.[4] These were for pickling, a process which required more capital—in the shape of salt and barrels—than did drying and which therefore was never undertaken by the fishermen themselves. Cod were mainly prepared in this way for the London market at Lent and curing was at its height in the early spring months.

By the last quarter of the eighteenth century herring fishing was pursued regularly and successfully on only one part of the east coast—the southern shore of the Firth of Forth. The old-established herring fishing of this coast consisted of a 'ground-drave' in which rectangular nets were held by stakes so as to present a face to the moving fish. This was on a small scale and lasted no more than one week in the year and occupying no more than 100 boats.[5] Then, in the early seventies, a 'float-drave' was established in the same general area between Dunbar and Eyemouth, occupying fewer boats but lasting for a much longer period

[1] *OSA*, XIII, 403.
[2] *OSA*, XIII, 402.
[3] BFS, Vol. 3/147, 'Report on the white fishing trade', 1787; Arbuthnot, *Historical Account of Peterhead*, 38; *OSA*, I, 471; V, 38, 278; XVI, 549.
[4] BFS, Vol. 3/147, 'Report on the white fishing trade', 1787.
[5] *Committee appointed to inquire into the State of the British Fisheries*, 1785, Reports from the House of Commons, 1803, X, 112.

of the year, from July to September.¹ This was simply local application
of the existing usual way of catching herring by drift net. These nets
were rectangular in form and in use were attached to each other in an
extended chain or 'drift', which would be cast into the sea and hang as
a curtain with the upper edge a short distance below the surface of the
water. With one end held by the boat, the whole chain was allowed to
drift so that moving shoals became entangled. Fishing was conducted
in the hours of darkness and after one or perhaps more shots a night's
fishing would be complete. Drift netting could be carried on direct from
the decks of large vessels, but it was in open boats worked from the
shore that the east coast fishermen pursued the herring. In the particular
case of the fishings of the south Forth, float-drave boats would have ten
nets, each rather larger than those applied to the ground-drave.² The
boats went to deeper water than they did for the ground-drave, but it
was still a small-scale fishing with only a slight tendency towards
expansion.

The herring is a highly perishable fish and if not sold quickly in a
fresh state has to be given at least a preliminary curing within a few
hours of being taken from the water. Open boats, such as those being
used around the Firth of Forth lacked the means of curing on board and
invariably had to make land shortly after they had been fishing. At the
points of landing they needed quick access to the consumer or some
form of curing plant. On the Firth of Forth, however, there was the
biggest consumer market in Scotland and some of the daily landings
could be taken direct to Edinburgh and Leith.³ But fluctuations in the
size of the catch made it difficult to sell each day's landings in a relatively
steady local market for fresh fish, and much of the catch was normally
cured either by 'reddening' or by pickling. In the preparation of red
herring, the fish, after being cleaned and salted, would be transferred
to the 'red herring house' in which they were hung, for the period of
smoking, on spits arranged on wooden frames. For this a permanent
structure was needed which had a rigid maximum capacity. The period
of preparation and smoking was an extended one, lasting for over three
weeks, and during that time the fish undergoing the process would fully

¹ David Loch, *Essays on the Trade, Commerce, Manufactures and Fisheries of
Scotland*, 3 Vols. (Edinburgh, 1778), II, 250; BFS, Vol., 4/619 'Remarks by Robt
Melvill, settler at Ullapool, on the herring fishing of Scotland', n. d.
² George Pitcairne, *A Retrospective View of the Scots Fisheries* (Edinburgh,
1787), 67; *Committee appointed to inquire into the State of the British Herring
Fisheries*, 1798, Reports from the House of Commons, 1803, X, 340–1; BFS,
Vol. 4/614, 'Remarks by Robt Melvill, settler at Ullapool, on the herring fishing of
Scotland', n. d.
³ James Miller, *The History of Dunbar* (Dunbar, 1830), 239.

occupy the available facilities. Pickling, on the other hand, could be conducted in the open air and the only equipment it needed were containers for the fish as they awaited the hands of the gutter, a store of salt and the barrels in which the fish were to be packed. Capacity to produce could be swiftly expanded simply by buying in more stores and hiring more labour. Herring, in being pickled, would first of all be thrown into a large rectangular trough (the 'farlin') and roughly mixed with a sprinkling of salt: sometimes, without further ado, they would then be packed, along with an additional and more considerable mixture of salt, in barrels which were then allowed to stand for a period of days with one end open to the atmosphere. But a higher standard of curing required the gutting of the fish before it was packed. Gutting was the most laborious part of the curing operation, and two women were needed as gutters to keep one packer supplied. Whether the fish were gutted or ungutted, the barrels would lie open at one end for up to a fortnight after the first packing; then, with some of the brine run off, the barrels would be filled to the top with more fish and closed. The process was then finished.

The fishermen making landings on the coasts of the Lothians and of Berwickshire had the advantage of a big nearby market in which part of their catch could be sold fresh—and therefore at good prices—but a considerable proportion was cured either as red herring or as pickled (white) herring. The fluctuating surplus was taken mainly by the curer for pickling as the growth of the float-drave tended to increase the importance of white as compared with red and fresh herring. But even by 1800, herring fishing employed no more than 200 boats for a few weeks of the year. It was still of small importance as compared not only with other types of fishing but also with some of the herring fishings that had occurred round the east coast in the past.

North of the Forth, herring fishing had either always been intermittent or (by 1790) had declined from the levels of success of the past. The weakness of the small boat for herring fishing was its limited range and its need to return to a base at which preparation had been made to handle the catch. The drift-net could be the instrument of great and easily-won catches when it hung in the way of a moving shoal, but the shoals which collected around the eastern coastline in the late summer were somewhat changeable in their habits of movement. It was food-supply—the plankton drifting on the surface—that dictated this movement but its location and sufficiency were determined by complicated and interacting influences of temperature, wind and current. Larger vessels, fitted out for curing the catch as it was hauled aboard and therefore capable of free movement, might adjust to changes in the location of the feeding grounds but small boats, closely tied to a fixed shore base, had to depend on the shoals entering narrowly defined areas of water, in no case more than a few miles from the shore. Thus a shore-based fishing, conducted in small boats, was liable to rise and fall with the changes in the movements of the shoals. Fife had once had a considerable herring fishing but through the eighteenth century it fluctuated around very low levels. There had, too, been fleeting occurrences on other parts of the coast of localized fishing opportunities, but only at unpredictable times and places. It was never easy to exploit sudden concentrations of fish on one part of the coast, for local markets generally were limited and it took time to summon the means of curing even when it was of the simplest.[1] Around the east coast most of the intermittent appearances, inshore, of herring led merely to the landing of a stock of fish that was almost unsaleable or, at best, to successful catching and curing of herring for a few years—only to be followed by the disappearance of the shoals and the decline of the fishing. Off the Banffshire coast, for example, herring were sometimes seen, but were not caught for lack of a market; on another occasion the large-scale landing of herring reduced price to a mere token.[2]

III

In the late eighteenth century a boat of the largest type used on the east coast cost about £25 and might be expected to last, at the most, for ten years[3] while the cost of providing a man's share in the lines would be about £2. 10s.[4] In some villages a class of smaller boats would also be kept, involving in each case an expense of £10 to be shared among four or five members. Sometimes landlords—the men on whose land the villages were sited—would give help in the purchase of the larger type of boat. Thus in Banffshire:

in consideration of receiving a specified rent annually, the proprietor allows £11 to each crew to purchase a new boat which is understood to last seven years, called here the long run. Then a mutual contract is entered into between the proprietor and the crew, wherein he engages to secure them in the property of the boat; and they bind themselves to serve in it, and to pay their rent during the term of 7 years. If the boat is judged unfit for sea before the end of the lease, and application is made for a new one, a deduction is made for every deficient year of the boat's run to the extent of £1.15 which goes in

[1] Souter, *Banff*, 297; *Fish Trades Gazette*, 3 Apr 1886.
[2] Souter, *Banff*, 297.
[3] BFS, Vol. 3/146, 'Observations regarding fishers and fishing villages', 1787; James Donaldson, *General View of the County of Banff* (Edinburgh, 1794), 6.
[4] *OSA*, VII, 205.

part of the £11 for another boat. In different towns the rent is different. The average rent of each boat is £5.3.3 and 6 dried cod or ling.[1]

Even when the landlord was supplying a substantial proportion of the capital, the members of crew were to be regarded as owners. One Banff-shire landlord engaged men 'to secure them in the property of the boat' and, while they were contracted 'to serve in it for 7 years', their operations were, apparently, entirely free except for the payment of a fixed rent. This form of control was somewhat more extreme than any found elsewhere and, south of Aberdeen, landlords do not appear on the picture at all. When, in the south, the crew was unable to provide all the necessary capital, it was a capitalist who was called in, being given a share in the fishing in return for a loan. Along the Berwickshire coast, for example, landsmen would contribute to the cost of the boat and would take a share in its proceeds; and the general principle even with these intrusions of outside capital was for the crew to provide the bulk of, if not all, the capital required for operation.[2]

Given that a crew generally contained a group of at least three members who had put money towards equipment, there remained several ways in which the power, the financial responsibility and the profits might be divided among them. The normal rule was of the equality of members of each crew with each man contributing equally to costs and drawing equally from profits but there were exceptions and qualifications. Usually a boat would carry a boy of under eighteen years of age who would contribute less in equipment—only half as many lines, for example—and would draw less in profit. Correspondingly there might be a tiny hierarchy of command within the crew. Among the Fraserburgh crews, for example, there were four ordinary members of crew, one foreman and a skipper.[3] But such exceptions to the rule of equality were unimportant. The 'boys' after all were on their way to full membership of the crew and no genuine social gap existed between them and the other members, each of whom would have gone through the same stage of training. As often as not, the boy was linked in the crew with his father or his brother. Similarly, the division between skipper and crew does not seem to have been based on differences of social position or of the provision of capital to the venture; the skipper drew little, if anything, more than the other individual members.

More seriously divisive were differences in the contribution made to the expenses of operation which designated one man as wealthier than

---

[1] *OSA*, XIII, 402.   [2] Miller, *History of Dunbar*, 237.
[3] John Cranna, *Fraserburgh: Past and Present* (Aberdeen, 1914), 56.

the other and which led to variations in earning. The extreme case was that of the landsmen who were hired for the season to make up the crews in Berwickshire.[1] They had no share in the boat or its equipment and were virtual wage-earners. In fact, the difference of standing between them and the other members of the crew might well be that between the true fishermen and the outsider called upon to help at the due season, for they came from outside the fishing community and their position on the lowest rung did not disturb the substantial equality of the others (the 'true' fisher members of the crew). It was, in any case, only at certain seasons that they were engaged; at other times crews would be made up, in equality with each other, by 'true' fishermen. One division between such true fishermen arose with differences in the individual contribution to expenses. Nearly always, where fishermen sailed together, they would contribute equally to the set of lines, each man providing a 'line' of given length and given number of hooks. Although the standard in each respect might differ from one community to the next, the local equality of provision seems to have been invariable and seems to have existed as much to symbolize equal participation as for convenience in sharing expenses. Such arrangements applied both to great- and small-lines. Every adult male and certainly every head of family in a fishing community would have his share in a boat along with between two and six other members and sometimes, where two different grades of boat were used, he would be a full member in two different crew groups. In the provision of the boat itself, a greater expense, there was less equality. Every boat was substantially, if not wholly, provided by the totality of the working members or by a group within them. Sometimes all would provide precisely similar amounts, but sometimes one or two members would provide the boat while the rest contributed only their appropriate shares in the lines. Differences of contribution meant differences of share in the proceeds but this applied only on top of the basic and symbolic equality of sharing among all participating crew members. The amount that was shared on the basis of the differing contribution to expense was small compared to that which was divided in virtue of simple participation in the activity of fishing, and the inequalities of income which emerged from different capital entitlements were fairly small. For example, in a crew of six members one-seventh of the profits might go to 'the boat' (that is, to the man or to the group which had supplied this most substantial item) and the remainder would be divided equally between

---

[1] Miller, *History of Dunbar*, 237; Patrick Lindsay, *The Interest of Scotland* (Edinburgh, 1733), 193.

all the members; if, as an extreme case, one man had provided the boat he would take two shares from the total proceeds while the remainder would take one share each.[1]

Each fishing village around the east coast of Scotland was undoubtedly an individual, with its own way of doing. Yet the uniformities covering the whole array were clearer than the differences. All were composed of families with the one similar occupation, pursuing a common sequence of operations through the year, with the season for one type of fishing following another in precisely similar fashion for each family. Normally, too, all families would participate on the basis of a settled division of labour in which the man would fish and the women support the operation of a particular boat in various ways. Each village would maintain this pattern by taking each cohort of young lads for training and ultimate absorption in the crews and by each slightly younger cohort of females becoming the marriage partners of the men in their first years as full participants in the boats. None of the communities would easily absorb outsiders whether male or female. Every male would move to his full standing as the member of a crew, sharing more or less equally with the other members of the crew. Most of the capital needs for fishing were met by the general body of the fishermen; and every man, certainly every family, would provide more or less equal amounts to that end. As a consequence they would take broadly equal shares from the process of fishing. Some crews beyond doubt would be more successful than others, and some individuals might have slightly more property than others; but these were random differences which did not cumulate into systematic and long-standing differences between definite classes. The fishing community was almost class-less within itself.

Life in an east coast fishing community was toilsome, hard and occasionally of great danger. The boats, frail open craft, were taken daily into a sea of swiftly changing weather which at its worst could be savage; once at sea, they had to work back to a coast where most of the landing areas were without protected entry and towards which an onshore wind could drive with shattering force; even where there were harbours, they could be entered only at certain times of the tide, and then dangerously if the wind was in the wrong quarter. The journey to the distant cod-banks—a sojourn which might extend over days— meant exposure in the most cramped conditions with only cold provender for the crew. There were, too, the recurring tasks of launching

[1] Headrick, *Angus*, 97; Cranna, *Fraserburgh*, 56; Lindsay, *The Interest of Scotland*, 193; *OSA*, VI, 3.

and hauling boats above high-water mark. If it was a life dangerous and hard for the men, it was also endlessly toilsome for the women. Some villages lay so close to the water that the carrying of fish and gear was minimized; but in other cases the cottages would be separated from the beach by cliff paths, and fish and lines would have to be carried for some distance over difficult terrain. There was the continual and monotonous, if not hard, work of baiting lines. The drying of cod and ling was a process of considerable labour over a period of weeks, with the fish having to be frequently laid out, turned over and piled. Finally the selling of the fish in the locality might involve the fisher-women in tramping miles with their burdens.

Yet the economic result of this hardship, danger and toil was to give a reasonably steady return which, shared in almost equal proportions among the different families of the fishing community, meant a modest and secure individual income. 'All of them are Opulant and Comfortable, and live much beyond the common sett of Tenants and Labourers.'[1] Behind this comfort, which contrasted so strongly with the circumstances of their working lives, lay the controlling factors of catch and market, some of which favoured the type of fishing pursued around the east coast.

Long-line fishing tended to be fairly dependable in its returns. No series have been found covering the operations of east coast boats at this time but the analogies of such fishing at other times and areas suggest that in this type of fishing both the daily and the seasonal return varied within fairly narrow limits. It was a fishing, too, in which boats at a given ground would return with very similar results. Thus the physical return tended not to fluctuate unduly from year to year and tended to sustain the boats at the same level of success at any given time. Haddock fishing was somewhat less dependable but even here the variations occurred over the longer period and through considerable successions of years there would be a steady income accruing.

Conditions for selling the product were on the whole favourable. On the more northerly stretches of the coast—where in fact the greatest number of fishermen lived—the great fish were mainly cured by drying and were carried, by the fishermen themselves, to the markets of the south in which there were to be found competing buyers. Much of the smaller fish, and particularly the haddock, whether or not they had been lightly cured, would be sold in more local markets, sometimes being carried direct to the consumer by the fishermen's wives. In neither case, then, was much lost to intermediaries and, even on the most remote of

[1] BFS, Vol. 3/147, 'Report on the white fishing trade', 1786.

the coasts, there was no evident tendency for merchants or tacksmen to establish a control over the supply of particular boats or centres. Only one type of fish was in fact generally sold to a local merchant for curing; this was the small cod, which would be cured by pickling (that is, pickled and then packed in barrels).[1] Towards the end of the eighteenth century the trade in this product was increasing and the arrival of merchants among the north-eastern communities, seeking supplies to prepare for the market when it was at its height before Lent, tended to bring some extra money.[2] Certainly the incoming merchants were in no position to dictate terms to fishermen who had their established channels of marketing to which they might always return. For the fishermen of Fife and the coasts farther south, the main market was either among purely local populations—such as the miners of Fife—or among the people of Edinburgh and Leith, a market big enough to take supplies both from Fife and from East Lothian and Berwickshire as well as from the local fishermen of Newhaven; thus Fife crews would take their daily catch direct to the other side of the Firth. The scale of fishing at this date was not enough to threaten to glut this market.

[1] BFS, 306/1, 'Observations regarding fishers and fishing villages', 1800.
[2] BFS, Vol. 3/147, 'Report on the white fishing trade', 1787.

CHAPTER II

The Rise of Herring Fishing: Caithness, 1790–1815

I

IN THE eighteenth century fishing on the east coast had its ups and downs, its local advances and retreats, but in the main a well-established system continued, year in and year out, to provide a regular and full livelihood for a population which changed but slowly. Then, in the last decade, the speed of change began to quicken. New forms of fishing were explored, movements widened and population began to increase perceptibly. In fact, a virtually new fishing industry began to emerge, partly in the new fishing centres of Caithness, partly within communities at places long accustomed to fishing. The herring fishing, with roots running through old-established fishing communities where the basis and sole occupation had been line-fishing, was starting to rise to first position in Europe.

The most dramatic appearance of herring in the late eighteenth century occurred in the 1790s in the Firth of Forth. The herring were found in great shoals, close inshore, in 1794, and for a succession of years thereafter they returned in sufficient numbers for a big annual fishing and curing effort to be made. Vessels fitted out with curing stock arrived from the west coast by the Forth and Clyde Canal, while on shore about 100 curing yards were sited around the town of Burntisland. Small boats would converge from both sides of the Firth and ultimately 800 to 900 boats collected for this winter fishing.[1] It was possibly the biggest herring fishing effort that had ever been made, regularly over a number of years, on any small sector of the Scottish coastline. Around 1800, then, this seemed to be the likely base for the great herring fishery that had so long been planned for Scotland. Yet the shoals departed from the inshore waters as abruptly as they had appeared, and by 1805 all signs of the fishing had disappeared from the north shore of the Forth.

---

[1] *Committee on the State of the Herring Fisheries*, 1798, 360.

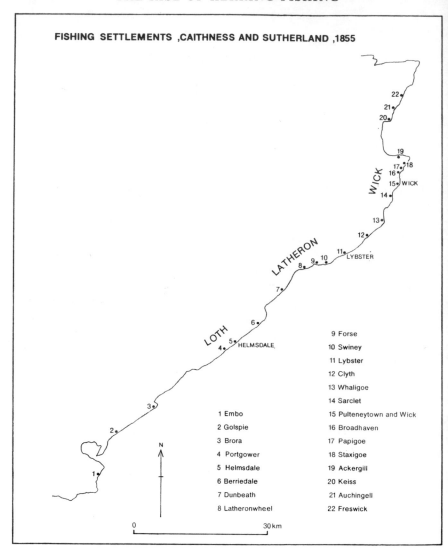

FISHING SETTLEMENTS ,CAITHNESS AND SUTHERLAND ,1855

1 Embo	9 Forse
2 Golspie	10 Swiney
3 Brora	11 Lybster
4 Portgower	12 Clyth
5 Helmsdale	13 Whaligoe
6 Berriedale	14 Sarclet
7 Dunbeath	15 Pulteneytown and Wick
8 Latheronwheel	16 Broadhaven
	17 Papigoe
	18 Staxigoe
	19 Ackergill
	20 Keiss
	21 Auchingell
	22 Freswick

0   30 km

Source: Parliamentary Papers, 1856, XVIII, Boat Account.

Meanwhile, the seeds of the successful herring industry that Scotland was to achieve in the nineteenth century were germinating in the remote extremity of the east coast, in Caithness, where there started an activity which was at first small in scale but which contained all the elements of method and organization necessary for success in the natural conditions of the east coast. This indeed was to provide the model for the all-embracing activity of the mid-nineteenth century which made the herring fishing as carried on by Scottish east coast fishermen the greatest in Europe.

Yet in Caithness, at the beginning, there was no sudden appearance of herring to attract immediately a flock of incoming boats. Rather, in somewhat tentative fashion, local crews were presented with the chance of selling their fish to incoming merchants, either by contract or by daily bargain 'on the beach' when the fish were landed. Fish merchants from the Firth of Forth, coming to the area intent on buying cod, first created the conditions under which the small local boats could profitably fit out for a limited period of herring fishing: Falls of Dunbar, for example, began in the seventies to buy herring to carry back to their red herring premises on the Firth of Forth.[1] They must have found a reasonable local supply because in the eighties they erected red herring houses in Caithness itself.[2] Thus by the late 1780s a system had been established by which about 100 boats fished seasonally—that is, from July to September—from Wick and Staxigoe to supply the merchants with fish for reddening and, later, for pickling. Local men, too, began to see their chance as curers and the market for fish became still greater and more certain. At first the landings seem to have been sold at a fluctuating daily price rather than on a seasonal contract. In 1789, there occurred the greatest fishing ever known on the Wick coast.[3] Four nights of it completely exhausted all salt and casks and the price fell from 8s. to about 2s. per cran. The implication is that for some years boats had been plying to sell to the curers at the expected price of 8s.

The fishing, however, survived the perils both of excessive and of deficient landings, and by the mid-1790s had increased to a steady annual effort involving about 200 boats and resulting in the curing of 10,000 barrels.[4] Pickled herring came to be a Caithness product of some reputation. 'Even when they are successful,' it was reported in 1789, 'quantities sent are far short of the demand for them both for home

[1] BFS, Vol. 3/82, 'Extracts from letters', 1787; 396, 'Notes by Lewis Mac-Culloch on the British herring fishery', 1789; OSA, X, 8.

[2] P. White, Observations upon the Present State of the Scotch Fisheries and the Improvement of the Interior Parts of the Highlands (Edinburgh, 1791), 32; Lewis MacCulloch, Observations on the Herring Fisheries upon the North and East Coast of Scotland (London, 1788), App., p. ix; BFS, Vol. 3/82, 'Extracts from letters', 1787.

[3] MacCulloch, Observations on the Scottish Fisheries, App., p. xvi.

[4] OSA, X, 10.

consumption and for exportation to the West Indies.'[1] The curers of
Caithness, in fact, 'were finding encouragement for more than they
could cure.'[2] If herring could be bought from the fishermen at 8s. per
cran there was indeed every encouragement for a venture in curing.
Barrels of cured herring could be sold, with no apparent difficulty, at
over 20s. each; to this had to be added the bounty of 2s. per barrel and,
if exported, a further 2s. 9d. Since curing costs, including the cost of
stock, amounted to 8s. per barrel at the most, there was a handsome
profit for even the least expert curer.[3]

Nor did the prospective curer need to lay out much capital on his
venture. Little material equipment or fixed capital was required and the
quick payment of the government bounty gave him some working
capital out of which he was able to meet his seasonal expenses even if
he had to wait some time for payment on his consignments. It is not
clear how far merchant firms were ready to buy from the curer immediately on delivery of the barrels and how far the curer had to consign
to the London or Greenock markets and then wait for the proceeds.
Clearly, the high prospective profit drew many small men—for example,
local Caithness tradesmen—into curing, along with the big southern
firms which had been the first to engage boats for herring fishing. By
1800 there were sixteen firms engaged in curing in Wick alone.[4] About
two-thirds of the catch was pickled and one-third 'reddened'.[5] The
average firm, then, would be turning out less than 500 barrels a year
and would give a seasonal employment to twenty or thirty workers.
The curers were mainly local men, of varied original trades, who saw
the chance of a good profit on modest capital in enterprises that need
not occupy their full time (for herring fishing was so definitely a seasonal
effort that it needed to occupy curers or fishermen for only three months
in the year).[6]

To secure the supplies they needed to profit from the situation where
the selling price of cured herring settled so much above their costs,
curers set up, or consolidated, a system of buying on contract.[7] The
price fluctuations which occurred under the sales system of the eighties

[1] BFS, Vol. 3/396, 'Notes by Lewis MacCulloch on the British herring fishery',
1789.
[2] MacCulloch, *Observations on the Scotch Fisheries*, 4–5.
[3] *John O'Groat Journal*, 12 June 1846.
[4] John M. Mitchell, *The Herring. Its Natural History and National Importance*
(Edinburgh, 1864), 35; *John O'Groat Journal*, 10 May 1844; BFS, Vol. 4/637,
'Report of Special Committee on the State of the Fisheries', n. d.
[5] Mitchell, *The Herring*, 321.   [6] Ibid.
[7] BFS, 283, 'Report by James Williamson on the state of the Society's estate
near Wick', 1805.

had sometimes hit the fishermen hard, and it seemed likely that a fuller effort would be secured by a system of guaranteed prices. Thus, in the nineties, curers stepped up their control of the fishing by making prior engagements with all boats. Each crew would be engaged by a particular curer to sell all their fish to him at a price which was fixed for the season.[1] Then the curer had the guarantee of securing the full supply from a given number of boats at a price which was known for the season. The engagements, however, were temporary; and in any year the price was determined by a process of bargaining influenced by the assessment of the prospect of the coming season and the number of crews seeking engagements in relation to the capital that was seeking a profit in curing. It was a type of arrangement that carried some risks for the curer. He was committed to pay a set price for his main raw material—fish—and, moreover, he had to accept and pay for an indefinite and unpredictable amount once the season had started; but the price which he would receive for his finished product was unknown and subject to many outside and uncontrollable influences.

But the new system did secure the curer an assured source of supply and his move to increase business found response among the fishermen as they fitted out more boats to engage in herring fishing. It is true that this fishing, still confined to an effort of little more than two months in the summer, cannot have provided anything like a living wage for the fishermen and was scarcely a full support for a true professional fishing population such as had been lacking in Caithness. The boats of the time usually averaged under 100 crans for a season's fishing and a crew of five would share among themselves less than £50, out of which the costs of equipment and the fixed expenses had to be met.[2] Such small incomes did elicit a fishing effort from among a population which had no existing class of full-time fishermen, but only because a seasonal herring fishing could be pursued as a by-employment of agriculture. Boats at first were small, costing no more than £30 each, and with a similar additional outlay on nets a crew were prepared for the fishing.[3] Thus groups of farmers, or even of cottagers, would pool their resources to fit out a boat for the herring fishing in the firm knowledge that they would make a worthwhile catch in nearby waters and that they would sell what they landed.

---

[1] *Committee on the Herring Fisheries*, 1798, 248; MacCulloch, *Observations on the Scotch Fisheries*, 5; *Fish Trades Gazette*, 3 Apr. 1886.

[2] Headrick, *Angus*, App., pp. 8–9.

[3] BFS, 361, 'Extracts from the rental, etc. of Pulteneytown', 1815.

The herring fleet was also made up of crews from the southern shore of the Moray Firth.[1] These were full-time fishermen from communities of long standing and of settled routines, responding to the chance of increasing their earnings by recourse to this new fishing on the distant side of the firth. Like the indigenous Caithness crews they were engaged by the curers for the season and would have to stay in Caithness away from their homes for the full period. Also contributing to the activity around the Caithness coast were 'busses', larger vessels fitted out on the west coast under the terms of the government bounty.

By 1800, then, the Caithness herring fishing had established itself as a modest but regular effort made by local and by stranger crews, with an orderly marketing system, a processing section which could deal with all landings, and the beginnings of a reputation for its product in distant markets. But it was still a fragile effort depending on shoals appearing within a few miles of one fairly short stretch of shore. The tiny boats had only scanty gear and were weak means of catching fish, for they could operate only at a short distance from shore and only in a dense shoal would the rudimentary gear give much return. The local boats were at most of 16-foot keel, although from the south there would come the larger cod boats of as much as 25-foot.[2] The smaller size of boat would only carry eight or ten nets, each of twelve fathoms (that is to say, each drift would extend to no more than 250 yards). The fishing itself seems to have been predominantly centred on the harbour of Wick although the crews were drawn from north and south along the coast.[3] This harbour was, in fact, little more than a river frontage without quays. The seas might roll dangerously along this front, entry could be hazardous in an onshore wind, and even in the calmest weather the tide controlled the period of access. A quite modest fleet of tiny boats caused serious congestion. Unloading could be difficult and all movement of stores and of products was cumbersome so that sometimes the traffic would almost congeal. Still, Wick was by 1800 an established herring exporting port accounting annually for up to 10,000 barrels, all of which would be removed by sea.[4] In addition to the chaotic effort of landing the output of some 200 boats, to be distributed among perhaps twenty yards on the shore, there would be

[1] BFS, 283, 'Report by James Williamson on the state of the Society's estate near Wick', 1805.
[2] BFS, 306/1, 'Observations regarding fishing and fishing villages', 1808; Donaldson, *Banff*, 5.
[3] *Committee on the Herring Fisheries*, 1798, 248; BFS, 361, 'Extracts from the rental, etc. of Pulteneytown', 1808.
[4] BFS, 361, 'Extracts from rental, etc. of Pulteneytown', 1808.

numerous sailings by larger vessels carrying away the product of these yards.

II

In 1808 there began a sudden steepening in the rate of increase and from the level of 214 boats in that year the fleet grew by large annual jumps till 1814 when there were 822 engaged in the herring fishing.[1] Such numbers were not, of course, entirely unprecedented for a herring fishing, but Caithness was to be different in continuing its increase over a long period: by the mid-twenties more than 1,000 boats would gather for its fishing. By then it made up a greater concentrated fishery than any that had been known in Scotland; moreover, this annual gathering of boats was to be generally sustained for another half century.

This great fleet, and the curers who served it, continued to operate in remarkably dangerous and awkward conditions. The congestion of Wick harbour and the evident prospect of drawing dues from increasing numbers of daily arrivals led the British Fisheries Society to build Pulteneytown harbour on the south side of the Wick river. The new works were completed in 1811, with accommodation for 400 boats.[2] But by then the fleet seeking a safe haven of operation had grown well beyond such numbers, and in any case the inconveniences and dangers of operating from either bank of the river were scarcely diminished by the new harbour. Access was still limited to a few hours in each tide, the danger of the easterly or north-easterly wind was still a constant threat and the congestion not only afflicted the daily efforts of unloading herring but also on occasion obstructed the general commerce of the port. For example, in 1819, some of the traffic from the outer coasts did not come as usual to Wick because the parties did not wish to expose 'themselves to the detention too often experienced by such vessels as load in the harbour'.[3] More and more, then, boats and curers were forced to use the most convenient—or least inconvenient—of the landing areas up and down the coast. Mostly these stations were coves around the base of the cliffs (to be entered through the breaks in the solid wall) of which most of the east Caithness coast is composed. There the crews found themselves in worse conditions than those who were still using the main harbours. The returning boats would have to search out the tiny breaks through the solid walls of the cliffs; the

[1] Ibid.
[2] BFS, 311/2, 5, 'Report by James Williamson, agent, on the state of Pulteneytown', 1809–12.
[3] BFS, 376/3, 'Reports about settlement at Pulteneytown', 1818–19; *Fish Trades Gazette*, 3 Apr. 1886.

crews had to haul their boats on the open beach; the curing proceeded on the cramped areas of flat ground around the landing places, exposed to the weather but never to be postponed; often the fishermen and workers would have to clamber up and down narrow funnels in the rocks between beach and dwelling. Stores would have to be unloaded by small boats shuttling to one of the larger vessels lying anchored off-shore and the cured herring would be taken by small boats round to Wick before being transferred to the holds in which they would be shipped to their destinations.[1] It was in such awkward and dangerous conditions that a considerable proportion of the fleet had to operate, driven out from Wick by the overcrowding of the harbour. In time, some improvements were made in the open creeks in which the fishing and curing had started. Clyth by 1840 had a 'neat little pier' and Lybster was so improved as to shelter boats.[2] Farther north, the creeks around Wick tended to be better provided than the Latheron stations.[3] Yet a pier did not make a fishing harbour; access might still only be possible over a few hours in each tide and once within shelter the boats would lie grounded and inert. Particularly in Latheron much of the fishing was still pursued from stations with no artificial protection; in the 1840s in that parish there was fishing at seven points, at only two of which there were piers, so that some 170 boats had to work exposed and unprotected.[4]

Large vessels fitted out for long sojourns at sea and equipped with all the apparatus of curing—the system on which the legislators had pinned their hopes—played little part in the new rising fishery. Indeed, boats which could be manhandled to be out of reach of the tide or which could survive in the congestion of Wick harbour were a necessity on this coast. To secure the full use of the limited tonnage, they had also to be open boats. Such craft had the additional advantage of cheapness, which allowed the fishermen to retain ownership and made profit a possibility when only fairly small takes might be expected. The boats were equipped only for the catching—not the curing—of fish, taken at no great distance from shore. Every boat worked to a scheme of departure late in the day for the grounds where the night's fishing would be effected with a return early on the succeeding morning. Almost all the herring were then taken over by the curers to be cured by pickling.

Thus open boats remained typical; they were in design the descendants of those types that had long been used for line fishing on the east

[1] BFS, 376/3, 'Reports about settlement at Pulteneytown', 1818–19.
[2] NSA, XV, Caithness, 105.   [3] NSA, XV, Caithness, 104.
[4] NSA, XV, Caithness, 102–3.

coast; they used one or two lug-sails on masts of about the same height as the length of the boat but also, if necessary, they would be propelled by oars. Yet the traditional designs were adapted, for there was a continuing tendency to increase the size of boats, as it soon became clear that the small scale of the fishing vessels was limiting the size of the catch and of the individual share. Bigger boats able to carry more nets would obviously give larger returns and added investment was evidently worthwhile even in boats that were to be used only for a short season. Thus it became common at each replacement to build bigger and more expensive boats and to add more nets to the drift. It was the great burst of investment between 1807 and 1813—in which the fleet was nearly quadrupled in size—that this tendency became clear. In 1808 the boats had been of the smallest size, of 15- to 16-foot keel, and were fitted with only eight or ten nets; by 1814 no boat of less than 18-foot keel was being used and some were of 24-foot, while the smallest drift being used was of eighteen nets, each of twelve fathoms.[1] So the process continued year after year. In 1817 'the fishermen had in many cases been enabled by the produce of the industry to replace the small boats previously used by new boats of much larger dimensions and to provide themselves with fishing materials of superior value'.[2] This was but a beginning and the improvements continued over the next two decades, even while tidal harbours and open creeks were still in use. By 1840 a boat of 30-foot keel costing £60 or £70 was standard, while nearly as much would be spent on the twenty to thirty-eight nets that went to make up its usual equipment.[3]

III

When by mid-century the Caithness fishing had reached its full development about 1,000 boats were being contributed to the assembled fleet by the people of the region, and this out of a society which thirty or forty years before had possessed only a handful of full-time fishermen.[4] Most of the local men owned the boats on which they worked. Wick itself along with the new settlement of Pulteneytown acquired a large population of fishermen, in part by immigration.[5] Some of these

---

[1] BFS, 361, 'Extracts from rental, etc. of Pulteneytown', 1815.
[2] FB Rep., *1817*, 4.   [3] *NSA*, XV, Caithness, 101.
[4] *NSA*, XV, Caithness, 153; *Select Committee appointed to inquire into the state of the Circulation of Promissory Notes under the value of £5 in Scotland and Ireland.* P.P. 1826–7, VI, 81.
[5] *NSA*, XV, Caithness, 144; Sutherland, 104. *Reports addressed to the Lords Commissioners of Her Majesty's Treasury in 1856, on the subject of the Fishery Board in Scotland*, P.P. 1857, XV, 22.

engaged in white as well as herring fishing and made a full livelihood from the one occupation in its different aspects, but many even in the town combined fishing with some quite different trade. In many of the landward areas of Caithness and eastern Sutherland the growth of a population with a major stake in fishing, both as to income and as to investment, had been no less notable. It was by a partial shift from agriculture that the numbers in the new occupation swelled. Farmers had been tempted into investing in boat and gear and into spending the full season of herring fishing at nearby stations. Often they continued to balance the two occupations in their annual programme but the income from fishing, flooding through the agricultural regions, started deeper reactions. Immigrants moved in, particularly from Sutherland, and the agricultural area along the coastal strip became severely congested.[1] In the parish of Latheron the population grew by 227 per cent between 1811 and 1841, and as the numbers swelled so were the farms broken down into smaller units to allow the combination of fishing with small-scale farming.[2] The crofter-fisherman was here the typical figure, although small groups of landless fishermen appeared around the small stations where the boats were kept and often worked. Crofter-fishermen were also found near Helmsdale, where new holdings, specifically designed for fishermen, were laid out on the moors by the planners of the Sutherland estates.[3]

While the major investors and most of the active participants in the fishing were local men, temporarily high levels of seasonal activity depended heavily on a massive concentration of temporary migrants. About one-third of the crews engaged were incomers, sometimes from other countries and frequently from the extreme south of Scotland; this would bring over 1,000 strangers, mainly from Fife and from the Lothian and Berwickshire coast.[4] Often such crews would make the long journey north year after year, each to some accustomed point of operation. Some also came with their boats from the west coast but the main contingent to arrive from the west was of hired men, arriving by foot on the chance of finding employment on the Caithness or visiting boats.[5] At least 3,000 such men would find employment for the two months of the season. The work in the curing yards was carried on partly by local women, wives or daughters of the fishermen, or by out-

[1] *NSA*, XV, Caithness, 41, 93.
[2] *Census, Great Britain*, 1851. Number of the Inhabitants, P.P. 1852–3, LXXXVI, Population Tables 1.
[3] Eric Richards, *The Leviathan of Wealth* (London, 1972), 205.
[4] *NSA*, XV, Caithness, 153; *S.C. on Promissory Notes*, 1826–7, 81.
[5] *NSA*, XV, Caithness, 101.

siders in search of seasonal jobs; some of the visiting fishermen, too, would bring their women-folk and these would look for work; but mainly, again, it was the migrant from the west coast who met the need. Wick, on which the bulk of the immigrants converged, more than doubled its population during the period of high activity.[1] From this seasonal influx the little town acquired some more permanent residents, as a few of the hired men chose to stay and become permanent fishermen. In its growth Wick became a melting-pot for the migrant people of the north and north-west.

The rising herring fishing in Caithness depended much on the enterprise of curing. With a negligible local market in fresh fish, fishermen depended solely upon the curers for purchase of their catch. The marketing of cured herring was inevitably long and complicated and the curer, at least in the first stages of the rising industry, had both to hold stocks of herring over protracted periods and to direct the distribution towards a distant market; and only skilfully processed herring would find a sale in the new markets that were necessary to a full-grown industry. The first men to undertake the curing and the sale of the Caithness herring were from the south. They had come seeking supplies of cod, had seen the local opportunities of herring fishing, had engaged local boats, and had ended by becoming the regular purchasers of part of the annual catch to be processed in their own yards. In 1805 there came a second wave of incomers when the Forth herring fisheries failed and the firms which had operated there for ten years turned to Caithness as the most hopeful area to replenish their business. But in the meantime local curers had appeared in increasing numbers; in 1800 the great bulk of the curers were local and this domination was never again challenged.[2] This was possible in part because curing (a trade of much risk) could be undertaken with little initial capital. From the local community came a sufficient body of men possessing accumulations of capital derived from various sources and ready to take their chance in the new industry. In the early 1820s there came the first big check to the industry, and with it 'a great reduction of the outfit by the withdrawing of sundry carpenters, shoemakers, druggist and others from the business'.[3] But any check was temporary. The Caithness fishing was solidly established. The ups and downs, inevitable with herring fishing, never cut into the solid basic level of business created each season. In the meantime, however, a herring fishery, for the moment of roughly equal scale but in the future of predominating importance, had arisen farther south, among

[1] *NSA*, XV, Caithness, 154.
[2] Mitchell, *The Herring*, 321.
[3] *John O'Groat Journal*, 12 June 1846.

the crowded fishing communities of the Moray Firth and east Aberdeen-shire.

Even by 1815 Caithness fishermen and curers had achieved much. For twenty-five years they had been operating a fishery of no little scale and at the end they were swiftly expanding it yet further. But perhaps just as important was the example it gave of how the herring stocks of the east coast of Scotland could be cheaply and reliably exploited. It provided a model that was capable of being moved to other parts of the Scottish coastline. In fact the means had been discovered by which the whole of a rapidly growing body of fishermen up and down the east coast could find profitable employment in at least a seasonal herring fishing. And, once they turned to herring fishing, much of their traditional way of life had to change. In 1815 there were mounting signs that the older communities of white fishers were beginning to take to the new herring fishing.

CHAPTER III

The Widening Sphere of Herring Fishing, 1815–1835

I

TILL 1815 the growing herring fishing had been virtually confined to Caithness, but then came a sudden widening to envelop a string of new bases. Within two years the harbours of the south shore of the Moray Firth were astir with the activities of herring fishing, pursued over the same period of the year as had been established for the herring fishing of Caithness; and when, by 1820, Peterhead had opened as a herring station, a fleet of over 2,000 boats would gather every year in the stations dotting the coast all the way from the northern tip of Caithness to the eastern coast of Buchan.[1]

In the new area of operation, comprising the southern shore of the Moray Firth and a short portion of the east Aberdeenshire coast, the catch (as in Caithness) was mostly cured by pickling, a task organized by local men drawn from varying occupations, who saw the opportunities of the new boom industry and engaged boats to fish from the small tidal harbours, usually serving the commerce of the region. There was no single predominating centre as was Wick within the Caithness complex. Fraserburgh and Peterhead each soon had over 200 boats but along the coast of the Moray Firth there were eight centres with local complements running from 30 to 100 boats.[2] Many of the villages which provided crews for this fishing had no harbours of their own; the men from them would move, for the season, to the centres of curing. Many crews, too, came from farther south—from Kincardineshire, from Fife, from the Lothians and from Berwickshire—to make up a total fleet which soon began to rival in size that concentrating in Caithness. Already, by the mid-1820s, at least 1,000 boats would gather for the summer fishing around the Moray Firth and Buchan coasts, to be added to well over 1,000 gathered at the same season at the northern

---

[1] FB Rep., *1819*, 473; James Thomson, *The Value and Importance of the Scottish Fisheries* (London, 1849), 20–33; *S.C. on Promissory Notes*, 1826, 81.

[2] Thomson, *Value of the Scottish Fisheries*, 20–33.

centres.[1] Landings in the region by 1821 were sufficient for the curing of 100,000 barrels.

In one vital way the new herring area differed from Caithness. In and between the new stations there already dwelt a relatively big fishing population, fully specialized and professional by 1800. Some crews from this region had moved, seasonally, across to the Caithness shore as soon as the curers were offering engagements to boats there for the herring season, but only a minority would make this journey to a

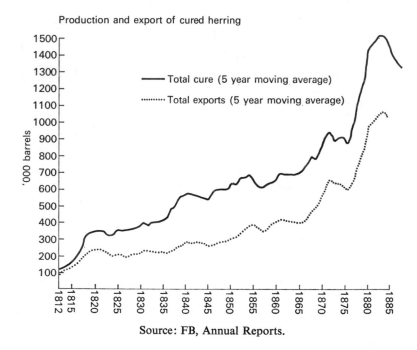

Production and export of cured herring

Source: FB, Annual Reports.

strange coast.[2] Thus, when herring fishing was mounted this close to the older villages the activity could be sustained without the major social changes—the very creation of new communities—that we have seen to the north. The immediate consequence of curers offering to engage boats around the Moray Firth and Buchan coasts was that almost the whole fishing population north of Peterhead was drawn into the summer herring fishing. Moreover, the existence of a herring fishing which was not so very distant from their homes attracted more and

[1] S.C. on Promissory Notes, 1826, 81; Thomson, Value of the Scottish Fisheries, 20–33.
[2] OSA, XIII, 403.

more of the crews from the villages south of Peterhead. Before long nearly every fishing community lying anywhere on the eastern Scottish coastline was sending some boats and crews to the summer herring fishing of the more northerly stretches. By the mid-thirties, then, at least seasonal and partial participation in herring fishing had made some permanent imprint on the old fishing communities of the east coast of Scotland.

The effect on incomes is not easy to trace. Presumably crews turned away from the traditional forms of fishing because there seemed to be the chance of higher incomes in the new. Yet the herring fishing was so uncertain in result that it may have been as much a gambling instinct as the likelihood of greater gain that turned men to it. In fact, the gains from herring fishing were only modestly higher than those from the fishings which were being forsaken for the new activity. Average catches oscillated from year to year, usually between 100 and 150 crans, with an occasional descent below 100 crans. Until 1850 the rate agreed between curers and fishermen generally stood at less than 10s. per cran, and the gross takings made by the average boat in the eight-week season would lie somewhere between £50 and £75.[1] The impact of the fluctuations on the fisherman, however, was increased by the fact that, whatever the gross return, owners of boats and gear had to meet fixed or nearly fixed expenses. Wages, payment for net ground, barking of nets, harbour dues, all would amount to at least £25.[2] Thus the amount available for division among share-fishermen fluctuated, quite unpredictably, between £25 and £50 with the chance of moving from one extreme to the other in successive seasons. On a boat shared by three owners, the individuals would take out anything between £8 and £16. As a weekly rate this implied a fluctuation upwards from the level of earning normal through the rest of the year. The gains of herring fishing, then, were on average modestly above those of other types of fishing, with the occasional year when they were considerably higher; but in most years they were little in advance and, very occasionally, hardly anything at all would result from the two months' effort. The hazards of uncertain catch were, moreover, greater than they might appear from the average figures. In any year there was a very wide difference of individual success, with the best-fished boats running up totals three or four times as high as the average. The prospect of the highest gains, however, were in fact delusive for the great majority. Generally, the majority of all boats would make something less than

[1] *NSA*, XIII, Banff, 236, 338; XIII, Elgin, 156, 209; XV, Caithness, 153.
[2] *Fraserburgh Advertiser*, 25 Dec. 1873; *John O'Groat Journal*, 21 June 1850.

the average for the district and the successful were an élite minority of boats which came out on top year after year. For the majority of crews the most likely outcome was a gain that was slightly above those to be made from other types of fishing. And to be set against such a hope there was the expense of buying and equipping boats which were often used only for the short season of herring fishing; the depreciation on a herring boat and its nets would be over £10 a year, representing, in a three-owner boat, a personal cost of more than £3.

II

The prospects or the results of the herring season were by the fourth decade of the nineteenth century very much the talk of the fishing press and, we may guess, of the fishing community. Perhaps it was the very uncertainty of the outcome that gave such discussion its fascination. Yet, in fact, while by 1830 the great majority of fishermen along the east coast were playing some part in this fishing, for very few did it occupy more than two months in the year. Even in Fife, where there was a winter fishing in addition to the summer fishing in northern parts, much less than half the year would be spent in herring fishing. The traditional 'white' fishings, carried on in much the traditional way, occupied most fishermen for most of the year and probably continued to provide the bulk of their income in much more certain fashion than did herring.

Even within the old-established fishings there were continual shifts of interest, changes of method and movements in catch and price. The haddock fishing, on the whole, was on the advance in the first half of the nineteenth century. Catches on the inshore grounds seem to have recovered from the crippling failures that afflicted some parts of the coast in the 1790s. In the North East, 1,000 fish might be landed as the result of a single trip to the inshore grounds.[1] Markets, too, were being extended. By 1830 the curing of haddock by smoking was common practice on many parts of the coast. There were several localized methods for curing. Smoking, over a peat fire, fish that had been split so as to lie flat was the method originated at Findon and widely copied; alternatively the fish might be similarly smoked over a fire of wood. Then there was the Auchmithie cure, a lighter form of smoking of fish only split sufficiently to allow gutting. 'Speldings' were the result of a completely different method of cure in which the fish were salted and dried in the sun. Most of these methods seem to have originated along the northern stretches of the coast but were widely copied wherever haddock fishing was carried on.

[1] *NSA*, XI, Forfarshire, 256.

Originally, too, the curing of haddocks had been a domestic occupation in which the wives of the fishermen used small private kilns attached to the house, such production being usually associated with the fish-wife acting also as retailer in the surrounding countryside or nearby town. The widening of the area in which haddocks were caught for curing was accompanied by the emergence of independent establishments for processing. Even in the north, where the older systems of domestic processing and local sales had their main roots, curers were coming to buy from the fishermen and complete processing in yards where the work of cleaning and of smoking was carried on by wage-earners. Thus, in Cove, near Aberdeen, the fishing people 'after gutting, cleaning, splitting, salting and smoking them with turf in a particular way, sell them in Aberdeen as Finnan haddocks'.[1] But in Bervie in 1837, 'in the course of last year, a few individuals formed themselves into a company for haddock curing, and made an engagement with the fishermen to supply them daily. They have hitherto been carrying on a very successful trade in dressing, smoking and barrelling the haddocks for distant markets, where they have found a great demand and ready sale for them.'[2] The first curers of haddock to come to Fife in the 1820s had the immediate object of selling their product in Glasgow, and by use of carts developed a trade which was to last till the coming of the railways increased its volume and profit.[3] Even in the north, where the smoking of haddocks was an old craft developed for a local market, factory production was symptomatic of a widening market. In the parish of Fetteresso, where haddock was the principal object of fishing and where peat-smoking had probably its original home, 'a much greater part, for some past years, has been smoked in houses prepared for the purpose, by wood'.[4] The product of these 'houses' was taken by cadgers to Forfar and Perth, by coach and sea to Edinburgh and Glasgow and by steamship to London.[5] The steamship did, indeed, play a considerable part in carrying this product so typical of the North East to relatively distant markets along the east coast.[6]

Even in the North East where herring fishing had taken such a strong grip, with a fleet expanded to the limit where there might be only two full-time fishermen in each crew, haddock seemed to be giving a return as great as herring and much bigger than cod and other great-line

[1] Ibid., 208–9.   [2] Ibid., 13.
[3] *Fife News*, 2 Jan. 1871; 12 Oct. 1872; 6 Apr. 1872.
[4] *NSA*, XI, Kincardine, 260.   [5] Ibid.
[6] *Royal Commission on Trawl Net and Beam Trawl Fishing*, P.P., 1884–5, XVI, App. B, Q. 958; *Commission on the Sea Fisheries of the United Kingdom, Minutes of Evidence*, Pt. II, P.P., 1866, XVIII, Q. 30830.

products. Thus, in Collieston, haddock fishing was said to give a return of about £1 a week to each man, a figure which fits fairly well with the estimate of £25 as the annual average of earnings from haddock in Boddam.[1] It is true that in this village the gross take of herring was valued at more than that of haddock; but the expenses also were higher and the net incomes coming to individual fishermen from haddock fishing were probably greater than that portion derived from herring fishing.

The great-line fishings—mainly for cod, ling, skate and halibut—were not so much affected by changes in method of curing, expansion of markets or fluctuations of yield. Cod continued to be caught on the same grounds, by the same methods, and to be cured mainly by drying although some were pickled. The market was wide, ranging from the Scottish lowlands to London and the Mediterranean, but so it had been in the eighteenth century. The apparent local slackening of effort in great-line fishing was due to the greater profit from haddock fishing rather than to inherent difficulties in the relatively declining activity. Thus, the average yearly earnings in Boddam from cod fishing were only £8 as compared with £25 from haddock fishing.[2] The cod and ling fishing, then, had generally retreated to take third place behind herring and haddock fishing as a provider of income. But in many parts it still continued to be vigorously pursued in the spring months.

III

In the traditional organization of the east coast fishing village, each man played his part, on much the same scale as his neighbour, in providing the means of fishing; in return he drew a reward and held a status very similar to his fellows. There were no extremes of wealth, no countervailing groups of employer and employed. In the nineteenth century new ways of fishing introduced obvious sources of inequality, with the splitting of crews into two groups (one of which acted as employers), with a growing range of reward and with an added strain on the weaker members in keeping up with the latest style of equipment. Yet, on the whole, the fishermen continued to contribute in broadly equal amounts to the equipment of the fleet. The typical fisherman was a man of increasing wealth but none became outstandingly wealthier than his fellows.

It was particularly the needs and opportunities of herring fishing that stimulated the effort to add to the value and scale of fishing equipment. The way to maximize income from herring fishing was for a partnership

[1] NSA, XII, Aberdeen, 279, 595.   [2] Ibid., 279.

of two or three men to acquire a boat and make up the crew with hired men. The fairly modest wages—usually amounting to less than £6 per head—that were paid to the hired men left much larger rewards for the owners than if the full crew held shares. The hired men were almost invariably outsiders, whether men from the west who came to the Aberdeenshire and Banffshire ports or 'half-dealsmen' whom the Fife crews gathered in their own agricultural hinterland.[1] A fishing community, then, which was heavily committed to herring fishing, would have one herring boat to every three or four fishermen and each man would hold a half or third share in such a boat. Numbers of fishermen were increasing, but for a period the herring fleet grew still faster.

Nor was this the end of the process. Many of the herring boats would in any case lie on the beach for the greater part of the year, for their larger size made them unsuitable for haddock fishing and an alternative stock of boats was kept which would be used for the inshore work. They would be manned by crews solely from the local community, men normally holding their boats under the eighteenth-century arrangement of equal shares. Thus, very often, a fisherman with a third or half share in a herring boat—which he might well use for only three months of the year—would also have a fourth or fifth share in a smaller haddock boat, together with the duty of providing his lines for the common equipment of the boat.

In all, then, many of the communities are found by 1830 to be in possession of stocks of boats several times larger than had been common in the eighteenth century. Thus the village of Boddam had three classes of boats, for herring, for cod and for haddock fishing, the boats of each type being used in rotation according to the season by the same general body of fishermen, who arranged themselves in different crew-groups to meet the needs of the particular fishing.[2] There were, in all, fifty-six boats owned by men drawn from eighty fishing families. This compared with a mere seven boats which had been used by forty-nine families in the 1790s, before herring fishing became so important.[3] It was more common perhaps to have two classes of boats, but even then there would be a great increase in the number of boats available to a fishing population which had itself greatly increased. Thus, in the villages of Portessie, Portknockie and Findochty the thirty-eight boats of 1790 had increased to 175 in 1830, although they were only of two types.[4] Population over the same period had increased from 583 to 1,634. These

[1] NSA, IX, Fife, 979; Fife Herald, 13 July 1871.
[2] NSA, XII, Aberdeen, 379–80.   [3] OSA, XVI, 551, 565.
[4] OSA, XIII, 401–2; NSA, XII, Banff., 261.

are villages in, and typical of, the herring fishing areas of the North East. But Fife also showed similar rates of accumulation and population increase: Cellardyke had 140 boats in 1837.[1] There were other areas which provided fewer crews for the herring fishing. Such herring boats might well be manned and owned entirely by the local fishermen who were fewer in relation to the community as a whole than when wage-earners were brought in. The property distribution in such a community had remained much like that of the eighteenth-century equivalent: thus the Kincardineshire villages of the thirties, still small, had apparently only one class of boats and these were fully manned by the local fishermen; six villages on this coast had in all thirty-four boats, owned and worked by 185 fishermen.[2]

Even by 1835 the individual boats had also increased in value and size under the impulse of the struggle for profit in herring fishing. Herring nets were bulky objects and a limit was set on the size of the drift by the capacity of the boat. Bigger boats could carry more nets without increase in the size of the crew; by increasing the size of their boats and by adding more nets to the drift fishermen raised their catching power relatively to the size of the crew. Thus from the time that fishermen were seriously involved in herring fishing the tendency was to strive always for bigger and bigger boats which could carry more and more nets. By the 1830s, boats of over 30-foot keel were being used for the herring fishing and the drift would typically consist of more than twenty nets.[3] Virtually all boats were undecked, usually with two lug sails and oars, but with many differences of design from one part of the coast to the other. While there were many minor local differences, there seem to have been two main types. The one was the boat characteristic of the Moray Firth which was descended from the Buckie boat of the eighteenth century with its heavily raked stem and stern. The other was based on the design originally typical of the Forth but carried north to Aberdeenshire and to Caithness; a boat in which the stem and stern were both vertical in profile. The increases in scale were achieved without much change in basic design.

Increased size and improved gear meant greater expense. By 1840 some of the new boats cost £100 each, although £60 or £70 was probably a more common figure for the herring boats.[4] Fitting a new boat with nets cost almost as much as the boat itself and, taking into account the fact that nets had to be more frequently replaced than boats, the expense on their account was probably the greater. Thus the fisher-

[1] NSA, IX, Fife, 979.   [2] NSA, XI, Kincardine, 208, 260.
[3] NSA, XV, Caithness, 101.   [4] Ibid.

man with a half or third share in a herring boat, being also responsible for the nets—and we have seen such a man to be typical of the communities given over to herring fishing—would have to put up between £40 and £80 simply to share in a fishing which occupied little more than two months in the summer. To this must be added his investment in the haddock boat and, possibly, where a different class of boat was used for cod fishing, in yet another boat of intermediate size. A reasonable estimate for this investment would be £20 (including the cost of lines). Thus in the villages where there had been heavy investment in increasing the capacity for herring fishing—the villages, for example, of Banffshire and of Aberdeenshire—the typical fishermen might have invested £100 in the various fishings in which he played a part. The Fife towns, also, had accumulated large stocks of boats but with their slightly different system of ownership—with one man as owner and the others providing gear—the wealth was presumably less evenly divided. In the other parts more moderate boat stocks are found, but there is no doubt that everywhere the fisherman was becoming more and more a man of capital.

The position of capitalist brought with it the implication of responsibility for replacing capital as it wore out. Boats had a life of ten years at the most and nets had to be replaced even more frequently, so that even to maintain their position fishermen would each have to find about £10 a year. Yet the better gear did not serve to increase catches over the years and the price per cran remained obstinately fixed around the 10s. level.[1] Thus the outcome of the herring fishing usually left the sharing fishermen with under £20 each for the season's work. This, added to not more than £40 for the rest of the year's work, left a small income indeed out of which to save £10.

The only source of aid for the fisherman was the curer and it seems that from early in the century curers would provide towards meeting the cost of new boats.[2] Yet this carried serious limitations on the freedom of action since while in debt to a curer a crew would have to fish for him alone and would be paid at a lower rate than the free crews. The evidence is that, until the late 1840s the crews everywhere managed to clear themselves of debt quite speedily, and that aid from a curer only slightly postponed the day when the fisherman would meet the full cost out of his pocket. The growth of the herring fishing and of the wealth of the fisherman was based on rigorous saving out of incomes which, in spite of greater fluctuation, were still usually very moderate.

[1] NSA, XIII, Banff, 262, 338; Elgin, 209.
[2] John O'Groat Journal, 29 Jan. 1845; 29 Feb. 1847.

Herring fishing had already by 1835 considerably altered the relation of fishing to the wider life of the village or community from which the fishermen were drawn. Fishers were still very much a group apart, marrying among themselves and receiving very few immigrants into their midst.[1] Yet the activity of fishing itself, together with the whole range of connected functions, tended to break out of the strict framework of village life. Fishing ceased to be based entirely on the homes from which the concerted efforts of whole families had performed every necessary function of preparation, catching, processing and even much of the selling. For one thing, herring fishing immediately drew many crews to distant centres of fishing. For two or three months whole families would move to the stations where the curers were offering terms to incoming crews. Sometimes—as with the men from the numerous villages of the Moray Firth and of Buchan which had no harbour and no local herring fishing—this meant a move of a very few miles but even so they seem to have shifted, with their families, for the whole season. For others there was a much longer migration; from Kincardineshire they went to Buchan, from Fife to the North East or to Caithness; and, with the decline of the summer fishing on the Berwickshire coast, from the Lothians and Berwickshire to the more northerly centres. Many villages, then, were completely deserted at the time of the summer herring fishing and probably a majority of all fishermen on the east coast were involved in some move.

Herring fishing's demand for ancillary labour was not contained within the framework of the family group as had been the older forms. For a period at the beginning of the century, nets would be made by the fishing families but the invention of net-making machinery about 1820 killed this household occupation. Nets were then purchased from the factories which were situated mainly in the Lothians and Fife.[2] The barking of nets, carried out before the beginning of the season, would occupy the women of the village before the boats departed for their fishing stations; but during the fishing itself the crew would operate without the direct aid of their women-folk. Nets had to be spread for drying but it was done either by the crew or by labourers hired for the purpose. The transport of fish from boat to yard was the responsibility of the curer and he would engage specialized carters for the purpose. The women on the other hand, finding employment within the yards as gutters and packers, did not directly support the crews in which their men-folk were organized. Rather they were hired as wage-earners by

[1] NSA, XIII, Banff, 257.
[2] David Bremner, The Industries of Scotland (Edinburgh, 1869), 312–20.

the curers to share impersonally in the work of a yard into which the catches of many boats might be discharged. They were paid largely on a piece-work basis and family earnings, swollen from such work, were further enmeshed in the varying fortunes of the herring fishing.

In spite of the impersonal assignment, for part of the year, of so many of the tasks that had been performed within the village (and purely by the folk of the village) there was no tendency to desert, or even to diminish in population, the sites where the fishing population of the later eighteenth century had pursued its localized activities. Villages grew in size, sometimes quite dramatically, but the population tended not to concentrate in the main centres of herring fishing to the detriment of the outlying communities. One reason for the continuing dispersal of the fishing population among relatively small units was the survival, and in some places the increase, of the haddock fishing through much of the year. As an inshore fishing this was generally pursued in boats of small type, and was directly based on the home whether or not there was an available harbour. All round the coast, small groups, usually of less than ten crews, would spend several months of the year at this inshore fishing. It was not only possible but also necessary to use the home as a base since fishing of the small-line type involved the daily baiting of thousands of hooks for every boat and each member of a crew depended on having his measure of lines baited by the family. Thus the system by which the women-folk directly supported the boat by work performed in the home was fully re-instated for the duration of the haddock fishing. Even for this fishing, however, there was some erosion in the self-sufficiency of the family group in carrying out all the operations connected with it, as the curing process became separated when, as we have seen, fish were often sold direct from the boat to the curers who ran independent units of factory type.[1] Sometimes the crews would enter into contracts with curers for the whole season, handing over the whole catch at a pre-arranged rate.[2]

Great-line fishing had never leaned so heavily upon the labour of the family workers ashore. Boats might be at sea for three or four days at a time and the lines, with far fewer hooks than the small-lines, would be baited on board; even when they were baited ashore it was by a lesser labour than was demanded by the small-lines. Curing by drying, how-ever, could be a demanding family task and it was much more slowly brought under the control of the curer than was the processing of

[1] Thomson, *Value of the Scottish Fisheries*, 65-6; *NSA*, XII, Aberdeen, 380; *Sea Fisheries Commission*, 1866, G.Q. 29799-800.
[2] Thomson, *Value of Scottish Fisheries*, 65-6.

haddock. Curers, nonetheless, did have an interest in the purchase of cod for pickling.

The whole work of the village, then, when it started again after the herring fishing, was a shifting blend between the older systems in which the whole family strove in preparation, in catching and in curing and the new, more specialized, activity in which the crew sold their catch to independent business groups.

IV

As we have seen, the great upswing in herring fishing—a concerted seasonal effort by virtually the whole east coast fishing population which brought together between 2,000 and 3,000 boats—owed little to the stimulus of price. At no time up to 1850 was there a continued upward trend in the price of cured herring and the 20s. a barrel which had been the normal price in the late eighteenth century was never exceeded by much or for long periods. The bounty of 2s. per barrel, raised to 4s. in 1815, did bring a fairly large proportionate return on each barrel completed and sold, and decisively increased the margins on which the curer could depend. With these additions to the basic price a rate could be paid for raw herrings which made it worthwhile for farmers in Caithness to equip for a herring fishing and for full-time fishermen farther south to turn from white to herring fishing in the summer months. And, when the bounties were removed in 1829, the curers were still able to pay enough out of the return from 20s. per cured barrel to keep the effort at full stretch. The industry, in fact, had found a technique of fishing and curing which allowed adequate incentive and reward for the different groups, given the average yields of the restricted inshore waters in which the fishing still had to be pursued.

Yet it was the indispensible pre-requisite of applying this technique that sales should expand, at least without serious drop in price, to allow disposal of the vastly increased cure that was brought to market. Throughout the eighteenth century, export markets were already of great importance to the herring industry. The main trade was in pickled herring and the main area of sale was the West Indies where this product was used in the feeding of slaves.[1] The focus was in London where West India merchants made up their cargoes for the planters but some herring also went from Glasgow and Greenock.[2] None, however, would be shipped direct from the fishing ports. Payment was made in six- or

---

[1] *Committee on the Herring Fisheries, 1798*, 103, 223, 243.

[2] *Committee on the Herring Fisheries, 1798*, 315; *Select Committee on the Salt Duties*, P.P. 1818, VI, 111; *Fish Trades Gazette*, 3 Apr. 1886.

three-month bills. In war-time the West Indian trade became rather irregular as shipments had to wait for convoy arrangements. Another fairly important market was found in Ireland to which, until the 1790s, there went something less than 10,000 barrels as compared with the 30,000 going to the West Indies.[1] The centre of this trade was the Clyde where merchants would buy for consignment. In the case of Ireland, however, it was common to ship more directly from the fishing centres. In Campbeltown, for example, merchants in a small way of business would fit out sloops for fishing and would use the same vessels to carry the cured catch to Ireland.[2] When, therefore, the northern and eastern ports had a surplus to sell, the curers would also ship directly to Ireland and, until at least the middle of the nineteenth century, Wick would send some of its cure direct to Ireland.[3] Whether his shipments were made for sale in Greenock or went direct to the Irish side the local merchant or curer thus had some period to wait before he would get the return on his cure.

In the 1790s, although tending to be irregular, exports to Ireland rose above the steadier West Indian quota; but the pattern which emerged after the disturbed trade of the war years was one in which Ireland stood slightly behind the West Indies as a market, with both taking much more herring from Scotland than they had before the wars. But the main change emerging during the war was the rise of the Continental market, taking it by 1820 to a position in which it rivalled, but stood slightly below, the other two. While the Continent remained still third as a market area, its rate of increase between 1810 and 1820 was by far the highest and there is little doubt that the towns of the Moray Firth and Buchan, rising so quickly as herring centres after 1815, were primarily stimulated by Continental demand.[4]

During the 1820s growth in all aspects of the herring industry was less remarkable than it had been before; exports to the Continent declined for a period, the West Indies took much the same at the end of the period as at the beginning, and sales to Ireland increased moderately.[5] Up to 1835, then, the rise of the trade with the Continent had been very important for the first period of rapid increase but the steadier rise in exports to Ireland, which continued till the 1830s, was the more persistent

---

[1] *Committee on the Herring Fisheries, 1798,* 278.

[2] Anne Rosemary Bigwood, 'The Campbeltown Buss Fishery' (Unpublished M.Litt. thesis, University of Aberdeen, 1972), 76.

[3] *John O'Groat Journal,* 5 Sept. 1845; FBR AF 36/68–77, Wick, Herring Exportation Books.

[4] FBR AF 24/23–8 Fraserburgh, Herring Exportation Books.

[5] FB Reps.

expansive influence. Yet this understates the importance of what had been happening in the trade with the Continent. Even stagnation in the 1820s kept in being a trade which was later to expand, first so as to fill the gap left by the decline in West Indian imports and then, after 1850, with exports to Ireland stationary or even slowly declining, so as to achieve an increase which left it completely dominant in the affairs of the industry. After 1850, then, the Continent became the main market, completely predominant in the end and capable of absorbing a very high proportion of an output which had itself grown many times over. This was a trade which demanded a new approach to the problems of curing and selling and the building of a new organization: developments in it were to be important not only for sustaining some growth in the industry up to the mid-thirties (the present period of discussion) but also in laying a foundation from which business could soar later in the century.

The Continental trade in cured herrings was large and old, even though for Scotland it might be new. All over the area east of the Rhine, and extending to those regions that could be reached by the great navigable rivers flowing into the Baltic, the pickled herring was a main foodstuff, whether taken as a delicacy or as a cheap basic sustenance. There was a potential for profit, then, in taking and in curing the herring that were to be caught in the North Sea and also in the trade by which they would be carried to consumers, many of whom lived deep in the Continental interior. For centuries both the catching and the carriage of the herring were virtually monopolized by the Dutch. Before the end of the eighteenth century, however, it became evident that the Dutch effort was declining even before competitors began to penetrate her traditional markets.[1] Because of the failure of the existing supplier, the Continental market in 1800 was ready for exploitation by whichever nations could find the resources and the organization. Scotland was a likely heir because she was so well placed to exploit the fishing grounds on which the Dutch had built their organization; but she was not the only country ready to break in, and the selling of herring on the Continent was a more complicated and testing operation than the traditional disposal in the West Indian and Irish markets. The custom of a discriminating market had to be achieved against strong competition.

It was well understood before the struggle started that the Dutch control had been exerted partly because of the superiority of their method of curing.[2] A considerable proportion of the cure offered by

---

[1] *Committee on Herring Fisheries, 1798*, 235, 303.
[2] *Fish Trades Gazette*, 11 June 1887.

any successor had to be sold in a market where quality as much as price was crucial. Particularly in Germany, herring was eaten as a relish rather than as a bulk foodstuff and it was difficult to sell herrings which were not cured by the Dutch method to which tastes were adapted.[1] More than one cargo from Scotland was rejected because of the poor manner of curing.[2] Of this the Fishery Board was well aware, and as soon as it took powers in 1807 it set itself to improve the manner of curing pursued in Scotland. Its success was quick and by 1810 Scottish exporters were finding that they could sell on the Continent.[3]

Curing performance was much affected, probably for the better, by the power and policy of the Fishery Board. This body was given a considerable staff of local officers and the whole Scottish coastline was divided into districts, each managed by at least one Fishery Officer. Armed with this staff, the Board set out to raise the standard of Scottish curing. The necessary mode was supplied by the Dutch and almost from its first year the Board started issuing regulations. Through its officers the Board administered the bounty (from 1786, 2s. for each barrel of cured herring) and, by laying down terms on which barrels of cured herring became eligible, its regulations were made effective. In fact, it remained legal to cure herring in any fashion, but only barrels conforming to the stated standards were considered eligible for the bounty. Such herring had to be gutted with a knife, had to be properly selected according to size and original quality, had to be packed with a specified quantity of salt in barrels of standard size and make, and had to stand in open barrels for at least fourteen days before being filled and closed.[4]

The central process was the gutting and packing. Invariably this would be performed by women grouped in teams (crews) of three, with two to gut and one to pack. Several crews would stand around the large rectangular container into which the fish were thrown on first arrival in the yard. The gutters prepared for the next stage both by gutting the fish—which was done in a single deft movement—and by dropping the gutted fish into containers which were differentiated according to grade. From there the packers would take the herring to complete barrels of designated grade. Whatever the overall size of the curing yard, the

[1] BFS, Vol. 3/395, 'Notes by Lewis MacCulloch on the British herring fishing', 1798.
[2] Committee on Herring Fisheries, 1798, 361.
[3] FB Rep., 1810, 204.
[3] Thomson, Value of the Scottish Fisheries, 107–9.

crew of three was the basic work unit.[1] The division into crews also gave a standard to determine the size of the labour force needed to serve the fleet; one crew of women would be hired for each boat engaged by a curer. The women were engaged for the whole season but, apart from the earnest money (arles), paid upon agreement, remuneration was mainly by the piece. A crew would perform the whole curing process for the barrels on which it had operated and was paid at a rate per barrel. The women's earnings depended upon their speed and skill and, in fact, they acquired great speed of hand in the continually repeated operation in which a fish was lifted from the farlin, was slit, the gut removed and the carcass passed to the packer.[2] Neither had the curer much difficulty in securing quality. Fishery Officers inspected the cure on a random sample basis: batches of barrels tested and found eligible for the bounty were branded, and for most curers this was a sufficient control; the brand, indeed, came to be the guarantee required to sell the barrels in most known markets. To secure this minimum quality for his cure, the curer had simply to withhold payment to the curing staff on barrels which were found by the Fishery Officer to be inadequately cured. Thus the whole curing operation came to be geared to the requirements of the Fishery Officer.

The rising reputation of the Scotch cured herring was undoubtedly furthered by this use of the 'brand'.[3] Until continental merchants and consumers could identify a Scottish method of curing the brand was, of course, useless; but once it came to be recognized that it signified that curing had been conducted in a particular manner then it took on a life of its own. The Scotch herring could begin to move freely along the channels of trade, bought and sold purely on the attached mark, and to reach the higher levels of the market. The penetration of the continental market, then, was ultimately based on the detailed controls exercised by Fishery Officers in a multitude of yards up and down the Scottish coastline and on the growing uniformity of method of the ordinary curers. At which period the brand was fully accepted as a sufficient proof of the nature of the contents of the barrel is not clear.

The Continental market demanded speed as well as quality. Cured herring was, in fact, a semi-perishable commodity and stocks depreciated in value if they were not sold within a year of the first catching of

[1] Thomson, *Value of the Scottish Fisheries,* 59–61.

[2] Thomson, *Value of the Scottish Fisheries,* 107–10; *Report on the Fishery Board,* 1857, 22; *Reports on and since the year 1848 on the subject of the Fishery Board in Scotland,* P.P., 1856, LIX, 7.

[3] Thomson, *Value of the Scottish Fisheries,* 113; FB Rep., *1815,* 4; *Report on the Fishery Board,* 1856, 403; *Report on the Fishery Board,* 1857, 50–1.

the fish. Thus it was vital that herring which had been caught in the course of the summer and early autumn should reach their final destinations before the transport system on which they moved—to the Baltic ports by sea and overland by river and canal—was closed by the winter frost. The filled barrels had to be transferred from the port of origin in Scotland to a point of retail which might be up to 800 miles in the Continental interior within about three months. The movement from the German ports on which the Continental end of the trade centred—Hamburg, Stettin, Danzig—was controlled by German merchants and such distribution was in any case well-established and presumably effective before the Scots began to advance into the trade. The important link that the Scots were themselves reponsible for forging was to reach at the very earliest possible moment one of the main Baltic centres; the earlier in the season the higher the price. The physical problem of transport was met by the direct and frequent movement of trading vessels between the Scottish fishing ports and the Baltic. In this the scale and concentration of the fishing were important. When a particular port was handling a sufficiently large output, vessels could be commissioned which would load entirely with herring; higher in the scale of local output the sailings could become so frequent that a curer needed to wait no more than a day or two for his herring to be on their way. A port which had reached a level of production of 3,000 or more barrels per week for export at the height of the season was in a position to maintain sailings several times in the week.

Exporting also brought financial problems: for example, meeting shipping and insurance charges, and holding stocks till they were sold on the Continent. Clearly, curers needed skilled agents who could look after specialized problems and more particularly could sell wisely in the complicated situations of a large market where prices fluctuated, often unpredictably, under many influences. In the early days of the Scottish herring trade with the Continent, transport and selling there seem to have been undertaken by venturesome Scots from the herring ports.[1] They could not, of course, shield the curers from price fluctuation, but they could offer specialized services and they might grant financial help to the small curer who had to meet maturing obligations before the herring were sold to the German merchants. They were referred to as commission agents so that the curer would appear to have retained ownership until sale had been effected on the Continent.

The Caithness curers had been comparatively slow to develop a large

[1] Thomson, *Value of the Scottish Fisheries*, 41; *Report on the Fishery Board*, 1857, 13; Cranna, *Fraserburgh*, 310.

interest in the Continental trade, which for many years remained secondary in their accounts to the direct trade with Ireland and to coastwise shipments.[1] But the centres farther south, from the time of their first emergence, were closely tied to developments on the Continent; the years when the fishing was established along the Moray Firth and in Buchan were years of the first great upward surge in shipments to northern Europe and the life of these ports never ceased to be overshadowed by the passing conditions and the more permanent circumstances of the trade across the North Sea. It meant that the fortunes of a fishing community were to a large degree tied to markets that could be capricious and unpredictable. The price of cured herring in Stettin and lesser centres on the Continent fluctuated from day to day and it was an index eagerly followed by the whole commercial community of the fishing ports.

The fortunes both of fisherman and of curer were tied to the price of cured herring, but in rather different fashion. For the fisherman what mattered was the price, expected or actual, at the time when the engagements were being made for the season; fluctuations thereafter did not immediately concern him, for his price was guaranteed. But whether curers made profit or loss, fared more or less well than they expected, did ultimately rebound on the fisherman in the keenness of the bidding for boats in the following season. For the curer, the whole prospect of profit or loss turned on small movements of price during the season. He was committed to paying a set price for his herring so that every slight fall in price when he was selling—that is, in the early autumn—brought with it the risk of serious loss.

Increasingly curers looked with anxiety at the movement of prices on the Continent as the main influence on their profits. The fluctuations were frequent, often sharp and always unpredictable. The influences going into the determination of price were many and even after the facts of catch were known, it was impossible to be sure that the price would automatically correct for the excesses or deficiencies in the Scottish supply. The Continent still took only a fraction of the Scottish output so that a large fishing was not necessarily followed by an equivalent increase in export eastwards. Scotland, too, was only one of the suppliers and the Norwegian fishing in particular had an influence at least as great as the Scottish. Added to this were those factors affecting demand, such as movements in income and in prices of competing or complementary commodities. In general, then, curers had to take prices as they came rather than make prudential adjustment before the

[1] Mitchell, *The Herring*, 9.

beginning of the season. The only indicators they had when they arranged their terms for the season were the level of stocks and the existing price on the Continent. Neither gave any real forecast of what would happen when the season got under way.

The home market for cured herring was also of importance. Pickled herring was in Scotland a cheap foodstuff, more commonly eaten than meat by the poorer classes of both countryside and town. The influx of Irish to the towns reinforced this dietary habit; in Glasgow, for example, herring was an important constituent of diet. The home trade was, of course, never recorded as directly as the main export trades but its course can be traced as a residual of the other figures, giving exports and total production. The figure is bound to be rather inaccurate. There was the possibility of herring being cured at home and so unrecorded by the Fishery Board officials; a considerable proportion of the west coast herring must have been used in this way. But on the east coast, where the bulk of the catch was being made, fishermen were generally bound to hand over all herring to the curers and once the herring had entered a curer's yard it was likely to be recorded since substantial bounties were paid on every properly cured barrel of herring: only where the inferior fish was less costly than the properly cured herring by a margin greater than the amount of the bounty would it be worth to carry on such unrecorded cure. In addition, herring which were sold fresh to the consumer escaped from the statistical record but in only two areas were such sales of any consequence. Firstly, some of the fish landed on the southern shore of the Firth of Forth went direct to the Edinburgh market as fresh fish. Secondly, much of the catch from Loch Fyne and other lochs of the Clyde region was despatched within the day to Glasgow. But along the more northerly sections of the east coast, which increasingly predominated over other areas, nearly all the landings went to the curer. Thus, the difference between the figures of export and of total cure can be taken as indicating the minimum amount sold on the home market, with the true total figure probably not much above the results of this calculation. It is possible, however, that the recording of the cure became more complete after the first years of the operation of the Fishery Board and that the indications of increased sales on the home market denote merely more complete statistical coverage. The apparent sharp growth of the home market between 1807 and 1812 may perhaps be disregarded. What is clear is that by 1815 home sales of cured herring were a major outlet and that, if we can regard the statistical coverage from then as being at least consistent, home sales tended to rise somewhat till the last quarter of the century.

CHAPTER IV

Herring Fishing Dominant:
the Driving Forces, 1835–1884

I

UNTIL THE 1830s herring fishing had grown and spread along the east coast without any apparent push from rising price. Rather, the industry grew by the slow erosion of traditional forms of fishing as more and more fishermen, stimulated to a new form of activity by curers, realized profit opportunities that had been inherent for many years. Plentiful fishings arising out of a natural condition that may well have been particularly favourable at the end of the eighteenth century, together with the granting of bounties on all cured herring, helped to create the prospect of profit; but, in the main, the movement was based on the spread of an innovation—a way of fishing and of curing which was cheap and effective to an extent that had not been recognized earlier in the eighteenth century. But from the mid-thirties rising prices emerge as the prime stimulant. Almost continuously over a thirty-year period more and more effort and ever better equipment went into herring fishing. As a result the activity, the wealth and the size of the fishing community of the east coast were all deeply changed.

The rise in the price of cured herring was irregular, but the trend was unmistakable. From 1835 to 1842 a sharp upswing stimulated fresh expansion, but a drop in 1842 left prices in a trough from which they were not to rise for about ten years. But in 1852 there began a rise which continued unchecked till 1857. From then prices moved irregularly, without definite trend, till 1875 when a sudden sharp jump led into steep yearly fluctuations, at an average level rather higher than for any earlier period of similar length. In 1884 the period of high prices came to an end, with a steep descent into a trough where they remained till the mid-1890s (Fig. 2).

These price movements were accompanied by a continuing increase in the catch and output of cured herrings. The total annual cure of the early sixties represented an increase of 10 per cent over the levels of ten years previously, while the further decennial increases to 1870–74, and then to 1880–84, were 32 and 56 per cent respectively. By 1880–84 the average annual production of cured herrings was 1,386,713 barrels as compared with the 598,639 of thirty years before and with the 380,286 of a similar five-year period in the early thirties. Even the low prices of the late eighties brought only a modest decline in output.[1]

The increase in exports was accomplished through a continual shifting in the main markets. In the late thirties exports to the West Indies fell to trivial totals, while Ireland was taking only slightly more than it had ten years previously.[2] The main increase was to the Continent and presumably in this rapidly increasing, but not yet dominant, sector is to be found the cause of the high prices prevailing till 1842. The later

Wick
Price per barrel of cured herring (Crown Brand Fulls)
Recorded during second week in September

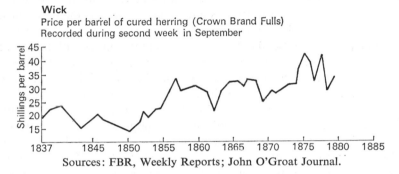

Sources: FBR, Weekly Reports; John O'Groat Journal.

forties were depressed because, with the West Indian market gone, there was at first difficulty and then collapse in the Irish market and little in compensation from the Continent. Both price and production fell during the decade with a decline in investment in new boats and actual withdrawal of some of the existing fleet from herring fishing. Then after 1850 the Continent took over as the main market for Scottish herrings, edging ahead in the fifties and advancing in the following decades to dominate completely in sales to both home and foreign markets. Annual exports to the Continent increased from 224,655 barrels in 1851–5 to 1,143,207 in 1881–5.

In 1886, in the first full account which included the disposal of the Scottish catch of herring in home as well as in foreign markets and the sale of fresh as well as of cured fish, exports took 69 per cent of the total output; 97 per cent of exports, amounting to 67 per cent of the total catch, went to the Continent.[3]

Increasing sales at higher prices suggest that the impulse to expand arose in part from conditions on the Continent. Even after 1884, when

[1] FB Reps.　　　[2] FB Reps.　　　[3] FB Rep., *1886*, App. A, v; C.

prices collapsed under the great flood of supplies being sent across the North Sea, a large proportion of the catch of an industry now inflexibly committed to a high output continued to be sent eastwards. It was only when the Continental market could absorb this undiminishing stream at remunerative prices that recovery would come.

The market area for herring remained very much that first penetrated and developed by the Dutch and then taken over by the rising fishing nations of Scotland, Norway and Sweden in the first half of the nineteenth century. In part Scotland's access to this market was controlled by fiscal policies.[1] Tariff barriers were greatly unequal as between one country and another, and were often operated selectively against different exporting countries. Further, the free trade movement of the middle decades of the century made little impression on these tariff walls. In spite of the trade treaty with France, for example, it never became possible for Scotland to compete in that country with her cured herring. In 1864, the duty levied on each barrel was still prohibitive in France, at 37s. to 40s. In Spain and Portugal the prohibition was more complete and Belgium also maintained a high duty. In effect, every country west of the Rhine was closed as a market. In the Rhineland itself the Dutch power was deployed not only in active competition, with a high-grade product and an advantageous transport situation, but also in the direct control of the river itself: the levy of 3s. on every barrel of herring for transit along the Rhine was a fairly serious impost. The countries of the Zollverein formed the main market and they were indeed the countries of reasonably low tariff; 3s. per barrel was charged —serious but by no means prohibitive, especially since competitors had to pay the same duty. Austria, on the other hand, potentially an important market, was practically ruled out till late in the century with a duty of 15s. per barrel. Farther east again the duty imposed by Russia appeared not to be too hampering but its effect was aggravated by two factors: firstly, the main rival, Norway, was subject to a much lighter impost, 1s. 6d. instead of 4s. 6d. a barrel; and, secondly, an annoying and wasteful inspection system seriously hampered the trader. This structure of duties dated back to the eighteenth century and was to be maintained with little change—except for some alleviation in Austria— till the end of the nineteenth. There is no doubt that it had an important effect on the shape and scale of trade. Yet it should be remembered that the countries effectively cut off by tariff barriers were those which in any case would have offered the smallest market.

Herring was sold on the Continent in distinct grades at widely

[1] Mitchell, *The Herring*, 270.

different price levels. The Dutch cure stayed supreme in quality but east of the Rhineland area it accounted for only a small and diminishing fraction of all herring sold; its price was too high for this herring to be an effective competitor in any market but the most exclusive.[1] The Scots had their greatest success slightly lower in the scale of quality and it is clear that they picked up the bulk of the custom of discriminating middle-class consumers; those who were able to pay something above the minimum for which the crudest quality of herring could be bought, the people who 'have always one dish at their tables—pickle herrings, and who would eat them—cut into slices—with bread and butter which they prefer to the best flavoured anchovies'.[2] The Scots' great rivals, the Norwegians, whether from laxity in the manner of curing or from the original nature of their herring, were never able to sell herring at the higher prices ruling for the first-grade product. Their great strength was low cost and in the market where cheapness was more important than quality there was keen competition between Norwegians and Scots.[3] The Swedish challenge had also been strong in this market, but had faded out when the fishing in their local waters failed. The Scots were able to maintain a reputation for quality in the market which demanded herring as a relish rather than a bulk foodstuff, and at the same time they made a challenge at the lower end of the market where they could only advance by selling cheaply and by clearly distinguishing between different types through branding; 'Fulls', the highest grade, sold at about 50 per cent above the lower grade.

The Scots sold their herring most easily in the geographically central sector of the market, in the belt running southwards from the coast between Hamburg and Stettin. Within this area were to be found a considerable number of consumers prepared to pay the higher price for the better grade of herring, and the Scotch product of the highest grade met their need at the expense of the Dutch. Even here, however, there was a market for herring among the poor who were looking for a cheap bulk foodstuff:

'As far as I know, white salted herrings when good are everywhere equally esteemed . . . and have been purchased because when cured according to the Dutch method they are very agreeable to the palate of the Germans and

[1] *Sea Fisheries Commission, 1866*, Q. 30749; *Select Committee on the Herring Branding System, and Revenue Appropriation*, P.P., 1881 IX, 212; *Report by a Deputation to the Continent to inquire into the new Branding Regulations*, P.P., 1890–1, LXIII, 1.

[2] BFS, Vol. 3/395, 'Notes by Lewis MacCulloch on the British herring fishing', 1789.

[3] *S.C. on Salt Duties, 1818,* 136.

partly because before they have risen to so great a price they afforded with a few potatoes not only a wholesome but a cheap meal.'[1]

As the Scottish trade with the Continent grew, it tended to centre upon Stettin. Total imports through that port increased markedly from 1825 onwards so that the import of an annual average of 32,566 barrels in the years 1825–9 had risen to one of 359,769 in 1874–8.[2] Every ten-year period saw its increase and only one of these steps was of less than 20 per cent. The trade through the port was dominated by Norway and Scotland. In terms of quantity imported, the balance swung from a Norwegian supremacy in the twenties and thirties to Scottish predominance in the late forties and fifties; in the seventies, however, the Norwegians came back to something like equality. These increasing quantities were taken at prices which were generally trending upwards. But the competition with Norway never disturbed the Scottish control of the quality market and the average price of Scottish herrings was always well above the Norwegian.

Stettin
Imports of cured herring from all sources and from Scotland

Source: S.C. on the Herring Brand, 1881, App. 10, pp. 224–5.

By 1850, therefore, the bulk of Scottish exports of herring were going to the Continent and progressively thereafter the proportion increased till only tiny amounts went elsewhere. At the same time, herring imports into the northern part of the Continent of Europe were dominated by two suppliers—Norway and Scotland. Yet, on the evidence of the size of the Scottish catch related to the autumn price of cured herring, there was no consistent link between the immediate post-seasonal price and the scale of the fishing in Scotland. Indeed, the consideration of imports

---

[1] *Committee on the Herring Fisheries, 1800*, 361.
[2] *S.C. on the Herring Brand, 1881*, App. 10, pp. 224–5.

from Norway indicates that the scale even of the combined fishings of the two countries provides only a small part of the explanation of price fluctuations. In fact, the chance influences bearing upon demand seem to have been strong enough, taken together, to obliterate the influence of short-term supply changes. Thus, in 1847, it was reported: 'how much the consumption of herrings depends upon a large, cheap supply of potatoes has been fully borne out by the late occurrences'.[1] On the other hand, a high price or a scarcity of meat and grain tended to increase the demand for herrings.[2] A more predictable and regular influence played its part in determining herring prices: to some extent the trade cycle, with its influence upon incomes, was a factor underlying the market. On the whole, however, it was the random and unpredictable influences that played the greater part. For example, it was said that in wine-growing areas a good vintage, leaving the vineyard workers with money to spend in the towns, would benefit the price of herring.[3]

II

The herring industry reacted sharply to buoyant markets and rising prices, and from the early fifties was pushed into a continuing expansion which, given the sharp short-term fluctuations of price, often carried substantial numbers to the perilous edge of commercial disaster. Yet actual disaster was averted till the mid-eighties, to give a thirty-year period in which growing effort and generally rising output were only briefly interrupted. The expansion of this one branch of fishing in its turn changed the daily life and the economic expectations of the general run of fishermen.

The business tone was at all times set by the curers; indeed it was only through them, the intermediaries between the fisherman and his market, that price influences were transmitted to the fishermen as a whole.

The first burst of growth of herring fishing had been secured by a multitude of small curing firms. Some of the men who undertook the shortlived venture of curing had risen from being coopers, the skilled wage-earners of the curing yard; but most were drawn from other, often diverse, sources, unconnected with herring fishing. Time created a more solid basis of long-established firms, each operating on a greater scale and tending to caution in business. Yet the trade always had a considerable section which was volatile in nature and was composed largely of

[1] *John O'Groat Journal*, 24 June 1847.
[2] *Fish Trades Gazette*, 1 Feb. 1890; *Northern Ensign*, 25 Mar. 1851.
[3] FB Rep., *1879*, p. xiv.

firms which were small, recent in origin and of little durability. Para-doxically, it was very often such firms that set the pace: in the engaging of boats no one could afford to be far short of the offers of the more adventurous—'The more skilled and intelligent curers are handicapped, even overriden, by men without proper experience, trading on borrowed capital'.[1]

Thus, in the Fraserburgh district, there were sixty curing firms at work in 1839 but by 1840 the figure was down to forty-seven.[2] Then the decline proceeded more gradually till 1860 when there were thirty-six. By 1870, however, the numbers had recovered slightly, to forty-one, a figure equalled in 1880. Since decline and then stabilization of numbers occurred within an industry of rising output the average size of firms increased over each ten-year period with the exception of that between 1850 and 1860. Yet the changes in scale were never sufficient to alter the character of curing as a trade of mainly small units. In 1880 there were three Fraserburgh yards which each accounted for more than 10,000 barrels, but even this represented the employment of only about 300 persons at the very peak of activity and a much lesser number through most of the year. The typical firm, even in 1880, would employ no more than 100 in the peak season.

There were some firms which developed interests in several ports around the coast, including the greatest of all, that of James Methuen.[3] Hailing from Burntisland, by the time of his death he had yards in nearly every main Scottish fishing centre. But the typical curer, while he might establish stations on the west coast and in Shetland, tended to centre his business firmly on the port which was his headquarters. Thus the bulk of the curing yards reported separately for each fishing district represented independent rather than linked units and the number of independent firms was not substantially less than the total of 590 yards reported, for example, in the returns for 1860.[4]

That the curing trade continued to be so volatile was the outcome partly of the inherent and largely unchanging nature of curing business, partly of the particular impact of the brand system—tending to en-courage the small firm and the new entrant—and partly of the policies of the banks and their agents.

Yards continued to be run as they had been in the early years of the century. Supplies of fish were acquired by curers through one of two possible systems of purchase. Along the southern shore of the Forth and

[1] *Fish Trades Gazette*, 25 Feb. 1888.
[2] FBR, AF 24/55–78, Fraserburgh, Herring Curing Books.
[3] *Fraserburgh Advertiser*, 5 Sept. 1862.   [4] FB Rep., *1860*, 18.

to some extent in Fife herring were sold by auction as landed. The curer was able to adjust his supplies flexibly; when demands other than curing raised prices he would buy little, when they fell he would buy more heavily but only sufficient to keep his yards going. He might be producing well under capacity for some of the time but on the other hand he did not need to buy when the price was too high to allow him any profit and he did not have to pay for unwanted herring. In all districts the produce of the winter herring fishing was sold under this system, but the herring which were landed in greatest abundance—namely the product of the summer fishing north of the Tay—were handled under the other system, that of engagements. It was the working of this latter system which determined how most curing was carried on and, indirectly, how the fishermen would be rewarded. In this case the curer would undertake to pay a pre-arranged rate for up to 250 crans landed from each boat and, very often, also a fixed bounty distributed at the time of engagement, which might well be several weeks before the beginning of the season.[1] For every boat on his list the curer would then hire a crew of three women to work as a team in the yard—with one individual to pack and two to gut. A small sum would be paid as 'arles' on engagement but, in general, the women would be remunerated at a rate, again negotiated from season to season, for every barrel which as a crew they packed, together with time-rates for their general work. Full settlement, both with the yard staff and with the fishing crews, was made by the curer after the end of the season.

Within the yards, with such dependence on the manual operations of these small groups, little development of technique was possible. Such limited change as was introduced stemmed from the edicts of the Fishery Board. Thus, the period of 'pining'—that is, the period for which barrels had to stand open as originally packed—was cut from fourteen to ten days; the specification of materials for the barrel was altered; the number of gradings for the brand was increased. But none of these rulings altered the way in which the main work of curing was performed. The Fishery Officers continued to play an important part in the industrial discipline of the yards, for they were responsible for the inspection of such barrels as were presented for the brand: implicitly they would see that the crews were performing their work properly. The speed of work, which was of vital importance to the curer, and the willingness on occasion to work many hours over-time were ensured by the system of piece-work.

[1] W. S. Miln, 'The Scotch east coast herring fishing', *Fisheries Exhibition Literature*, Vol. IX (London, 1884), 19-23.

While the work in the yard continued unchanged, there were important developments in the way in which cured herring were bought and sold among the dealers who stood between the fishermen and the (often distant) consumer. By the early 1850s the Scottish commission houses, which had handled the herring up to their sale in the German centres, were largely replaced by the German import firms extending their purchasing to the geographical heart of curing, in Scotland itself.[1] The new purchasing agents also brought with them funds which could be used, in the form of short-term loans, to support the seasonal operations of the Scottish curers.

There were, in fact, three main ways in which the curer might sell to the Continent and in all cases arrangements could be made for the curer to have his money in hand before the herring arrived there. Most commonly sales were made on the spot of barrels already completed. Almost every day of the season such sales could be made at prices which fluctuated from day to day, or even hour to hour, as information about the promise and achievement of the Scottish fishing, about the Norwegian effort, and about all the circumstances of price and supply among related commodities was passed to the market. Scottish curers 'by inducing the German dealers to come and buy on the spot transferred the value of the fish at once into their hands'.[2] The bill of lading and insurance documents drawn for most transactions were sufficient for the curer to obtain an advance drawn on a London bank with which the German purchaser had credit. On this he could obtain cash immediately, although presumably at a discount, a local bank being the intermediary for receiving the draft from London. The essential basis of such transactions was the brand.[3] In addition, the deepening interest of the German importers allowed some cash to be raised even before the season had commenced. Some of the cure could be sold before the fishing or the curing had even started, and on these forward contracts— which again were made possible because the brand had created a standardized product—the buyer would pay a certain proportion of the agreed price immediately, the balance being settled on delivery. In effect, therefore, the curer was getting a short-term loan which helped him to meet the expenses of fitting out for the season.[4] A third method

---

[1] *S.C. on Herring Brand, 1881.* QQ. 2774, 3001; *Northern Ensign,* 14 Sept. 1854; FB Rep., *1855,* 379.

[2] *N. Ensign,* 14 Sept. 1854. See also *N. Ensign,* 15 July 1851; *Sea Fisheries Commission, 1866,* Q. 31102; *S.C. On Herring Brand, 1881,* QQ. 84, 1198, 3085.

[3] *Sea Fisheries Commission, 1886,* Q. 37102.

[4] *Report on the Fishery Board, 1857,* 50–1; W. S. Miln, *An Exposure of the Position of the Scotch Herring Trade in 1885* (Aberdeen, 1886), 67–8.

was to consign to the Continent, the sales being made on commission by agents. In this case, the curer was venturing into a market where he might eventually be forced to sell at inopportune moments. Consignment, therefore, tended to be a last resort when prices at the Scottish end were low. Even so, there was much criticism of curers using this method of selling, on the grounds that they tended to be weak participants in a market which required the financial strength to hold back stocks till they could be sold at advantage; it was the men with considerable capital who were likely to come out best in such a market. In any case it is clear that only a small proportion of the Scottish cure was consigned to the Continent on account of the Scottish curers.[1] Even on consignment, however, the German firms which handled the sales in Europe on a commission basis were still willing to give advances amounting to a proportion of the likely price.

Some firms were of sufficient reputation, because of their size and history, to complete such transactions on the basis of their promises and of personal trade marks; their herring and the documents for their herring would pass from hand to hand, and money would be advanced by purchasers without the need of inspection (which for cured herring was an awkward and hampering process). But for most, and particularly for small and new firms, it was the official brand that allowed participation; the corollary was that the merest novice, having had his cure branded—as he could do on demand—could take full advantage of the system. 'We would be very slow to advance money upon a bill of lading if the brand was not upon the cargo.'[2] But with the brand, herrings—however produced—might pass into central Europe: 'The bill of lading of a cargo of full-Crown herrings is transferable like a banker's draft from one party to another in the interior of Germany.'[3] Thus, the herring of any curing firm, great or small, could pass to central or eastern Europe on the basis of a documentary proof and would at each stage command the going price of full Crown brands. Equally the brand, or the promise to obtain the brand, unlocked the credit of the German firms.

These credit arrangements fitted in with a trade which in any case needed cash-in-hand for only a small proportion of the value of turnover. Many of the curer's expenses did not have to be met till after the greater part of the product had been sold, often against bills of exchange

[1] *S.C. on Herring Brand, 1881,* QQ. 1198, 2404.
[2] *Sea Fisheries Commission, 1866,* Q. 31102. See also *S.C. on Herring Brand, 1881,* Q. 3654.
[3] *Sea Fisheries Commission, 1886,* Q. 26957.

that could immediately be discounted. Boats, it is true, had to be bountied ahead of the season with immediate cash and a small sum had to be paid as arles to the yard staff, but salt and barrels could be paid by a three-months' bill not falling due, say, till October; settlement for raw fish—the main cost—would only be made after the end of the season, and many of the labour costs in the yard were only met by a deferred payment. Thus a curer might engage his boats at 20s. per cran plus a bounty of £50. If the boats made an average catch for the season of 200 crans he would then be paying 5s. per cran in advance of the season; and the rest of the payment for the fish could await the settlement in October, when he would have cash in hand to the value of the barrels as completed. Additional costs of curing each barrel amounted to about 8s., mostly in the shape of the cost of the barrel itself; and, except for the cooper's wages, were paid for by bills which might not mature until the autumn. In all, at very most 20 per cent of the cost of curing would have to be met before the curer was able to realize some cash from the sale of the season's product.

Thus a curer needed to put up only a smallish proportion of his expenses before the funds coming in from his sales—accruing promptly because of the credit arrangements based on the brand—became available to meet the large residue on which the payment could be delayed. The gap in time between the outlay which had to be made early in the season and an effective realization of income was also short. If this gap could be bridged by borrowing, only a modicum of cash in hand was needed for a venture in curing; and operations as a whole could be extended almost without limit. Imported capital, we have seen, played a part here but even more decisive was the support given by the local banks. The usual form of bank loan was the cash credit, granted on a short-term basis. And no material security was needed, loans being made 'on personal credit, general standing, trustworthiness'.[1] Nor did this creditworthiness necessarily include any prior experience of curing, or indeed of business.

A party starts curing: he goes to a bank and states his views. The banker gives credit in the first place for wages to make barrels, perhaps to pay for staves, also bounties to boats and gutters' arles—in fact to pay all current expenses. When the fishing draws on, and herrings are cured and fit for the market, if consigned, the banker receives the bill of lading. If, after all the returns are got in, there is a profit, the banker receives his interest and charges for money lent.[2]

[1] S.C. on Herring Brand, 1881, Q. 2085. See also Q. 3118; Fish Trades Gazette 28 July 1888.
[2] Fish Trades Gazette, 24 Jan. 1885.

These credit policies allowed curers of all types to press their business on a flimsy basis of personal credit and undoubtedly they sometimes tempted curers into a headlong rush to secure the services of the available boats; rates paid to fishermen would then tend to move up towards the limit set by the ruling price of cured herring—which in turn created a precarious position, since the hopes about the market might well be falsified at the critical period when the main sales were being made.

Not all curers took their business to the maximum of their possible available funds. The discounting of bills meant taking a lower effective return on each barrel: where a curer made his own barrels he would incur running expenses throughout the year. Many curers extended their financial interests in other ways than by the operations strictly necessary to running a yard, as when they gave credit to fishermen for the purchase of boats—a form of transaction which was rewarding to the curer not only because of the interest he might glean but also because it bound the boat-owners to him as their creditor as long as the loan was outstanding. Yet there was always the tendency for even the conservative curers to be dragged along, at least in terms of the prices they had to offer in order to bind their crews.

It was in giving aid to the struggling newcomer, however, that bank support was most important. Loans might be given against the most tenuous promises and men of no capital, with no business experience and no training in the techniques of curing, might be aided; one case is reported of a boy of sixteen, without capital, receiving a loan to start in business.[1] And the support given to this element in the curing trade did not end with the lending of money. The bank might undertake some of the detailed business operations on behalf of the inexperienced: the bank or its agent would get the bill of lading, sell the herring, transfer the draft and deliver the money into the curer's hands.[2] Essentially, the banker hoped in this for profit in one of two ways. There was the commission he obtained on transactions completed as agent for a German purchaser and there was the chance of profit by speculation.[3] But in any case the agent had both the urge and the power to increase the curer's turnover; he could then secure control of the herring to make his profit. Banking authorities frowned on, but could not prevent, such private enterprise: 'The great bulk of Banker's herring, salt and store commission business, although tolerated is not encouraged by the banking

---

[1] *Peterhead Sentinel*, 13 Dec. 1887.
[2] *Fish Trades Gazette*, 29 Sept. 1888.
[3] *S.C. on Herring Brand, 1881*, Q. 3123; W. S. Miln, *Exposure of the Herring Trade*, 76.

institutions'.[1] Altogether, the self-interest of the local agents—using their power of lending and their position of consultants to gain personal control of a substantial slice of the output of cured herrings—was a strong influence furthering the drive to extend credit.

The inflation of business, by the pumping in of credit and the persistent recruitment of the untried, brought its own special dangers into a trade which was always tricky and beset with failures. Even the routine task of running a curing yard carried considerable risks. When, as with most curing businesses, the fish were bought under the engagement system, it was never possible to ensure that the price at which the cure was sold would be sufficient to cover the costs of making it. This was because the levels of costs were fixed before the season started without possibility of adjustment or of altering the size of the output, whereas, except for the small amount that was sold forward, the selling was effected at a price which fluctuated with little possibility of prediction. Thus a fall in price between June and September might mean that there was a loss on every barrel completed; but even then the curer was bound to carry on with his curing since he was committed to pay for, and therefore to use, all the herring that could be landed by the boats under the terms of their engagements to him. This very rigidity tended to make price fluctuation worse since there could be no corrective action, no slowing down in effort, when the market was showing signs of glut; the fishing would grind on remorselessly bringing more and more supply to a market where selling was already difficult. A curer, then, had to survive by skilful buying and selling.

In the 1870s and early 1880s these forces of expansion steadily took a stronger grip and the restraining hands weakened. High prices, increasing exports and plentiful credit brought in men fresh to the trade and caused even the old and conservative to abandon caution. As always, many of the curers had passed through training as coopers and were strongly committed to the occupation; most of the older and bigger firms were probably run by men of this stamp. But there were also others, a group probably increased in the time of expansion, who were drawn to curing because of the apparent chance of quick profit and described as 'enthusiastic and sanguine men of straw'.[2] They were drawn from a variety of occupations. On one list were to be found 'solicitors, grocers, one . . . shoemaker . . . even a grave digger'.[3] Such men might spend only two months in the year as curers. Trying a venture in curing in the

---

[1] Miln, *Exposure of the Herring Trade*, 78.
[2] *Fish Trades Gazette*, 22 Jan. 1887.
[3] *S.C. on Herring Brand, 1881*, Q. 591.

hope of quick profit, they would often manage the business by rule of thumb, keeping no business record other than a day book. They might well survive only by the help of a bank agent who would also be able to take profit from acting as agent for a German import firm.

Statistically, it appears that the newly-emerged firms always accounted for much of the business done. A tiny number of firms which had been in existence since the earliest days survived through to the later years of the century and these firms tended to be the largest. Thus, in the Fraserburgh district, in 1880 only three of the total forty-one firms had been in existence more than thirty years, but they accounted for 19 per cent of output.[1] Yet in the same year the proportion of the business which was produced by the young firms (of less than five years of age) was, at 37 per cent, greater than it had been in 1870 or in 1850. The firms of 1880 were considerably larger than they had been ten or thirty years earlier, but this was because of the quick growth of the still numerous newly-created firms as much as because of the great size of the old-established.

The result of business youthfulness and inexperience was a tendency to drive dealings to the limits of safety. It took only a small number, determinedly pressing their business, to bid up the rates paid for herring in a way that had to be followed by everyone. Added to a banking system eager to expand credit in general and, in the case of particular agents, following a personal interest in a big output of cured herrings, this caused great pressure to expand herring fishings. The credit pumped in by the banks undoubtedly helped to secure a big increase in output: firstly, it encouraged curers to bid for the services of boats and to support a level of prices which made a venture at the herring fishing irresistible; and, secondly, it eased, directly or indirectly, the provision of funds for equipment. For long the results seemed to justify the application of the forced draught and a rising market took the increased output without much strain.

But annually a very risky position was created. It is certainly true that curers tended to advance to dangerous positions, burdened by debts and trading on a precarious margin; for heavy debts were incurred as part of the normal seasonal operation. When James McCombie of Peterhead went bankrupt in 1888 he had liabilities of £86,000 of which £73,662 represented an unsecured debt to the North of Scotland Bank.[2] More typical of the failures, perhaps, was George Davidson of

[1] FBR, AF 24/55–78, Fraserburgh, Herring Curing Books.
[2] *Peterhead Sentinel*, 11 May 1888.

Fraserburgh who owed £3,922 to the North of Scotland Bank and had assets of only £553.[1] If it is less easy to glimpse the affairs of the firms which survived, all the evidence indicates that it was normal—or had become normal by the late seventies—to fit out for the season largely with borrowed money.

Fraserburgh

Rate paid per cran of uncured herring

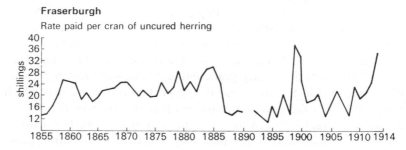

Sources: *Daily Free Press*, 8 September 1890; FB, Annual Reports.

Such debts were dangerous because their repayment depended on the highly uncertain movement of prices during the season, particularly in August and early September when the bulk of the cure was being sold. The trade was normally so organized that even a small drop in price over this period would bring many firms into difficulty. In the sixties and seventies, curers generally contracted for their fish at rates equivalent to between 20*s*. and 25*s*. per cran.[2] The cost of curing was, as we have noted, normally about 8*s*. per barrel, and if the price of cured herring dropped to below 32*s*. (including the inferior as well as the top-grade cure) many firms might be in difficulties. Yet 32*s*. was in fact a very high price even for top-grade (full) herrings before 1860. Nevertheless until 1884 there was no bout of extensive failures and, while 1870 was a nervous year and some curers did not meet their obligations in full to the fishermen, recovery came quickly and the seventies proved even more prosperous—with high prices all round—than had been the sixties. The reporter who in 1870 said that 'the chances have been so long in his [the curer's] favour that he has got to be overconfident as well as foolhardy' can scarcely have guessed that another fourteen favourable years were to pass before the overconfident were to be caught by recession.[3]

[1] Ibid., 25 Sept. 1888.   [2] *Daily Free Press*, 8 Sept. 1890.
[3] *Peterhead Sentinel*, 3 May 1880.

III

One limit to the growth of herring fishing was the lack of adequate harbours. The pattern of operation which had been established in Caithness in the late eighteenth century, with each boat engaged in a regular cycle of a nightly sally to the fishing grounds followed by a morning landing, was greatly hampered when boats had to wait for the tide before landing and leaving and where unloading meant a competitive struggle within small and congested areas. Yet the east coast was ill-provided with either the natural creeks or the full-scale artificial harbours which would allow fishing operations to flow easily. Both the general scale of the fishing and the initial siting of the fishing stations depended on whatever rudimentary harbour accommodation might be available although, at least in Caithness, the boats had sometimes to resort to the open beaches. But, when the possible profits of herring fishing became evident and when in particular the advantage of using large boats was proved, considerable effort was made to create new and to improve old harbours. Thus the basis of fishing was changed, the level and adequacy of accommodation was improved, and new social impetus was given at places which could combine the provision of harbour space with access to good herring grounds. The provision of harbour space was one factor, among others, in the efficiency of the fishing effort and in the determination of the areas and points of growth.

Even when only small boats were being used, the poverty of harbour accommodation can be seen to have obstructed the smooth pursuit of the fishing. Long delays in waiting for the tide had to be fitted into the twenty-four-hour cycle; boats had to be laboriously hauled above high water; and, at least in the creeks, there was little prospect of using larger and more efficient boats. Even more serious was the fact that the difficulties of re-entering tended to keep boats at home when there was any threat of bad weather. Thus, 'there are a great many creeks in which herrings are cured . . . where if it blows the least puff of wind the fishermen have to draw up their boats above high water mark'.[1] This uncertainty of operation might seriously diminish the results of the season's fishing, possibly by more than the proportionate loss of fishing time, because the best fishing opportunities arrived on only a few days of the season and, if any of these days were blank, there was a big loss of landings. The growth in boat numbers also soon created a gross

[1] *Extract from Captain Washington's (unfinished) Report on the Damage caused to Fishing Boats on the East Coast of Scotland by the Gale of 19th August, 1848,* P.P. 1849, LI, 59.

overcrowding at some of the centres, causing both inconvenience and danger. Peterhead, a harbour which could accommodate 351 boats in its inner basins, was in fact being used as base in 1846 by 437 vessels many of which were unloading from the seaward side of the pier.[1] How far the use of small harbours, with limited depth even at high water, held back the use of larger boats is difficult to assess. Certainly, boats of the size that became common in the seventies, requiring seven feet of water, could have been used in only one or two of the harbours available in 1820 or even in 1850; but for long the fishermen did not desire to build to these limits. In mid-century the inconvenience of operation in these crowded basins was undeniable; and it was worst of all, perhaps, at Wick.[2]

These manifest dangers and inconveniences, together with vistas of gain for the community or the individual providing for part of the rapidly growing fleet, spurred efforts to extend and improve harbour space. The projects of the middle decades of the nineteenth century were many; yet, till the late seventies, the additional provision scarcely matched the needs of a fleet which was increasing in numbers and in the size of its individual units. Much of the expenditure went piecemeal in small amounts, distributed among the many places ambitious to stake their claim for a share in the new fishery; where the schemes were of greater scale, difficulties of physical situation prevented any single centre securing all the attributes of a good fishing harbour. These attributes were indeed complex and expensive to realize:

'The requisites of a perfect fishing harbour are an entrance which will allow the boats to have full access and egress at all times of the tide; perfect shelter within the entrance; sufficient space for all the boats that frequent the place during the fishing season to lie together without crowding or jostling; enough depth of water in every part of the harbour to enable them to lie afloat at all times of the tide; and proper facilities for taking in their nets and gear, and landing their fish.'[3]

In the widespread effort to improve harbours, a small trickle of help came from public sources. The Fishery Board disbursed annually a small and rigidly limited sum of money for harbour improvement. Until 1850 the policy was deliberately one of spreading funds among places with no established position, in the hope of broadening the base and of distributing widespread the benefits of the herring fishing.[4] The

---

[1] *Captain Washington's Report, 1849*, p. v.
[2] *Sea Fisheries Commission, 1866*, Q. 30677–80.
[3] *Select Committee on Harbour Accommodation*, P.P. 1883, XIV, Q. 1342.
[4] *S.C. on Harbours of Refuge*, P.P. 1857, *X*, Q. 3137; *Sea Fisheries Commission, 1866*, Q. 26035; *S.C. on Harbour Accommodation, 1883*, Q. 2341.

result was that each scheme was largely ineffectual in meeting the needs of a fleet of bigger boats demanding ever more massive works. Only six east coast ports, all of them small, received aid from the Fishery Board's fund; and none became a fishing centre of any consequence.[1] After 1860 the policy changed and the still limited resources of the Board were concentrated on a few large schemes.[2] Notable and important was the financing of the Union Harbour at Anstruther, completed in 1877 at a total cost of £80,347.[3] This gave a large but somewhat shallow harbour that was usable, with some limits on the period when the entrance could be passed, by the biggest boats of the time. It was ironical, perhaps, that by the time Anstruther had its first-rate harbour there remained no summer fishery for herring in the area and the Fife boats generally moved north for the season. But the new harbour was important for other types of fishing—for the winter fishing for herring, for great-lining and for haddock fishing with large boats: it became the focal point for much of the year for the general fishing effort of the East Neuk of Fife. The effort to improve Dunbar harbour had a less happy outcome.[4] It took £35,582 of the Fishery Board's money, but the problems of easy and safe entry were not solved; from here as from Fife the herring fishing had departed by the time the work was complete but at Dunbar it did not sufficiently nourish efforts in the rest of the year to create an important centre of fishing.[5]

A second source of finance for harbour-building was the private land-lord, who might see the chance both of drawing revenue and of creating higher levels of income for the people of his estates from an active harbour used as a private possession. In the first three-quarters of the nineteenth century, several new harbours were built and many older ones improved by the private efforts of landlords. Some of the new ones became active centres of the second rank. Helmsdale in Sutherland and, farther south, Burghead and Macduff are examples from early in the century, with Boddam and Port-Errol coming later.[6] But the greatest of the private harbours was undoubtedly the Cluny Harbour of Buckie. This centre, with its long tradition of being foremost among fishing

[1] *S.C. on Harbour Accommodation, 1883*, QQ. 1259, 2341; *S.C. on Harbour Accommodation*, P.P., 1884, XII, p. viii.
[2] *S.C. on Harbour Accommodation, 1883*, App. 8; QQ. 2278, 2281; *Sea Fisheries Commission, 1866*, Q. 26130.
[3] *S.C. on Harbour Accommodation, 1883*, App. 8; Q. 2503.
[4] *S.C. on Harbour Accommodation, 1883*, App. 8; Q. 2549.
[5] *S.C. on Harbour Accommodation, 1883*, QQ. 2503, 2549, 2554.
[6] *S.C. on Harbour Accommodation, 1883*, Q. 3280; *S.C. on Harbour Accommodation, 1884*, QQ. 923, 1127, 2526, 2612; Thomson, *Value of the Scottish Fisheries*, 31.

communities and with its tendency to participate vigorously in all the developments of the nineteenth century, had strangely lacked any harbour of its own. In 1845, the fishermen themselves built a pier to enclose a small basin of less than two acres, but it was not till 1858 that a true harbour was opened.[1] This, the harbour of Nether Buckie, was even then small and tidal.[2] The full provision of a harbour for the town had to wait until 1877 when the Cluny Harbour was completed.[3] This was financed by John Gordon of Cluny at a cost of £60,000 and it was made to the highest standards of a fishing harbour, open at all states of the tide and capable of accommodating afloat at all times the largest contemporary types of fishing boat.[4] It was not large, and, with an irony similar to that which afflicted Anstruther and Dunbar, by the time of its completion there was no large fleet of boats eager to use it for the summer fishing. As a year-round base for varied local fishings and as a home for a fleet which might go to other parts for active fishing it was, however, efficient and important.

It was only by borrowing on the public capital market that the large amounts necessary to a large-scale and well-equipped fishing harbour could be raised. This was the way taken by those harbour authorities which ultimately were to be able to provide space and good conditions of fishing for the bulk of the herring fleet. Money was borrowed in adequate amounts for those harbours which were managed or owned by companies or trusts or municipalities and after 1861 their way was made easier by the Harbours and Passing Tolls Act which permitted borrowing at low rates of interest from the Public Works Commissioners.[5] It is significant, however, that such loans could only be made against the security of revenue from dues; and to prove such security it was necessary to have an existing harbour with considerable traffic. The Act in fact reinforced the judgement of the self-acting capital market that only those who were at least moderately successful before improvement could have the benefit of further considerable funds. It was no accident, then, that the main developments tended to concentrate at centres which were already the main commercial harbours in 1800. The great bulk of the spending was concentrated at Wick, Peterhead, Fraserburgh, and Aberdeen although Lossiemouth, Stonehaven and Arbroath had lesser schemes.

[1] *Captain Washington's Report, 1849*, xiii.
[2] *S.C. on Harbour Accommodation, 1883*, Q. 2267.
[3] *S.C. on Harbour Accommodation, 1883*, Q. 1147.
[4] *S.C. on Harbour Accommodation, 1883*, QQ. 1147, 2267.
[5] *S.C. on Harbour Accommodation, 1883*, Q. 1263; *S.C. on Harbour Accommodation, 1884*, p. v.

Ambitious schemes and large spending did not bring quick results. The improvement schemes at Wick provide a history of peculiar frustration. The harbour and the extension built by the British Fisheries Society on the south side of the River Wick, opened in 1811 and 1830 respectively, considerably increased space for boats but were far from solving the local problem of harbours.[1] Even the inner parts of the harbour could still be dangerously raked by a swell rolling along the river, entry could often be hazardous and there was extreme overcrowding in the available space. Furthermore, the efforts to create a large and safe harbour for the great number of boats that continued to gather there in summer were unavailing. In 1862 a scheme of improvement was accepted by which the government was to lend £60,000 in addition to £40,000 provided by the British Fisheries Society. In fact £132,000 was eventually spent, including contributions from surplus dues;[2] Yet it was all in vain, for great storms in 1870 and in 1880 swept away the new structure, and by 1884 Wick could be said to have had a harbour no better than it had been fifty years before—that is, a harbour grossly overcrowded and not very safe.[3] If the number of boats had decreased since 1850, they were now very much larger, being of up to 55-foot keel as compared with the 30-foot boats of mid-century. On returning to harbour, therefore, boats might have to lie off for hours, so that their herring might well deteriorate.

Fraserburgh and Peterhead also had a long struggle to create fishing harbours satisfactory in all respects. The first half of the century saw a succession of improvement schemes in both ports so that Fraserburgh had spent £50,000 by 1850 and Peterhead £100,000 by 1860. But the major schemes were still to come. Large-scale spending started at Fraserburgh with the so-called Balaclava project for a long and stout breakwater to enclose a large sheltered area with adequate depth of water.[4] It was not completed till 1873 when, in all, over £200,000 had been spent on the improvement of the port—a figure which has to be compared with the £9,000 the Fishery Board was allowed to spend annually on all its schemes. The result was a harbour of thirteen acres, capable of accommodating over 800 boats, allowing passage at all states of the tide and providing extensive deep-water berthing

---

[1] *S.C. on Harbours of Refuge, 1859*, App. 2, 311; *S.C. on Harbour Accommodation, 1883*, QQ. 4653, 4680; *Captain Washington's Report*, 1849, p. iii; BFS, 376/3. 'Reports on the settlement at Pulteneytown', 1818–9.
[2] *S.C. on Harbour Accommodation, 1883*, QQ. 4617, 4628–9.
[3] *S.C. on Harbour Accommodation, 1883*, p. xiv; QQ. 2530, 4628–30.
[4] *S.C. on Harbour Accommodation, 1883*, Q. 3321; *S.C. on Harbour Accommodation, 1884*, QQ. 35–51.

space.[1] Some deepening remained to be done but here was a harbour which could be used with convenience by the biggest boats of the day. Peterhead had a similar history, with success achieved on something like the same scale at about the same time.[2] Fishing could at last burst through the constraints which had kept its operations irregular and its boats below the most efficient level of size.

The combined force of the many harbour works, undertaken by landlords, by the Fishery Board and by trusts, companies and municipalities acting with borrowed money, tended at first to broaden the base from which herring fishing could be carried on. A measure of protection and convenience was provided at an increasing number of points, but at none of them was the facility good enough to allow untrammelled and safe fishing. By 1860 the herring fishing had spread widely among a large number of mainly small harbours, with only Wick, with all its deficiencies, taking more than one-tenth of the fishing fleet. Thus for the herring fishing of 1864, a fleet of 1,401 boats was gathered round the north-east corner.[3] Peterhead and Fraserburgh had the greatest aggregations, with 647 boats between them, while fifteen other ports shared the remainder at an average of just fifty boats each. At this time Fife also had an active summer herring fishing, its fleet being distributed among the cluster of old harbours in the East Neuk. Finally, Eyemouth and Dunbar and their neighbouring but lesser stations supported their own fleet. Not one of these harbours up and down the coast had water deep enough to allow boats to enter and leave at all states of the tide.

This pattern was considerably changed in the seventies because of the unequalled facilities now provided at Peterhead and Fraserburgh and by the emergence of Aberdeen as a main herring fishing port. By 1879, these three between them housed 1,917 boats for the summer fishing and consequently well over half the fishing fleet had fully adequate fishing facilities.[4] But the provision of good harbour space was not the only influence redirecting the effort of the herring fleet, as can be seen at Anstruther and Buckie which had good harbours but no longer had a summer herring fishing of any consequence. In general, the concentration of herring fishing on the main ports in the seventies left a long string of harbours, which had been painfully constructed or improved during the preceding seventy years, now devoid of the type of activity that had been the main object of their improvement. The effort of building harbours had not, however, been entirely wasted, for they might

[1] S.C. on Harbour Accommodation, 1883, QQ. 3287, 3319.
[2] S.C. on Harbour Accommodation, 1884, QQ. 848–9.
[3] Banffshire Journal, 16 Sept. 1864.   [4] Peterhead Sentinel, 10 Sept. 1879.

have other uses than housing a herring fleet. Even if great-line fishing tended to be based only on the deep-water harbours there was the haddock fishing, an operation covering many months, often carried out in small boats which could fish more regularly from a safe haven. And the big boats which ran to the larger centres for their herring fishing were periodically brought home and beached. By the 1880s the majority of east coast fishing communities had some type of harbour but there were one or two important clusters that still had nothing at all.[1] The group of villages south-east of Fraserburgh, containing 540 fishermen and 150 first-class boats, had not even a pier. They were, it is true, sufficiently close to the port of Fraserburgh to keep their large boats at the big harbour, but they still operated a small-boat fishing for haddocks from their own unprotected shores. The villages of the Kincardineshire coast, too, were poorly provided; some had small piers but some had no artificial protection at all. Generally, the communities without harbours were isolated points among the greater number that had. Caithness had seen much building of piers to serve its coves but sheer physical decay of these structures had left some of the communities, which were in any case declining as fishing centres, with no useful harbours. The Moray Firth coast, on the other hand, now had fourteen harbours among sixteen communities. Of these only Buckie had a deepwater harbour, seven only had some water at low tide but offered entry and egress only for limited periods, and the remainder were entirely dry for some part of the tidal period. From Fife southwards virtually every fishing community had its artificial harbour but only Anstruther had safe entrance at all states of the tide.

[1] FBR, Reports on Harbours, 1886, 1891.

CHAPTER V

The East Coast Fisherman, 1835–1884

I

THROUGHOUT THE middle decades of the nineteenth century the fishing population of the east coast was growing, to all appearances steadily. It was by the growth of existing villages and communities rather than by a spread of population into new areas that the increase occurred; but at the same time there was little tendency for the larger of the existing communities, or indeed for the active ports and commercial centres, to grow at the expense of the smaller. Of the seventeen villages picked out in the census reports of 1841 and 1871, only one showed a decreasing population; nine had gains of more than 50 per cent and three of more than 100 per cent.[1] Overall the number of resident fishermen grew from 17,659 in 1854 to 26,319 in 1880.[2] The larger number of the latter date was spread through very much the same number of communities as had existed at the earlier, so that fishermen were now residing in units which were appreciably larger than at mid-century and very much larger than at the beginning of the century. By 1886 over half the fishermen lived in communities containing 200 or more men of that occupation, whereas in 1800 the biggest of the villages had less than 100 fishermen; at the later date nearly a third were in units of 300 or more.[3]

The villages, thus remaining as numerous as ever, showed the traditional and unchanging characteristics of the fishing community—characteristics which marked off the fishermen and their families as a group apart. Even as late as the census of 1891 the communities of the east coast, apart from Caithness, were still composed solidly of men who had been born in the same parish as that in which they resided.[4] A few fishermen's wives might have come from other areas but nearly always they would have been born in parishes with some fishing community; the likelihood is that the women-folk were still almost exclu-

[1] *Census, Scotland, 1841*, P.P. 1843, XXII, Enumerators' Abstracts, Pt. II; *Census of Scotland, 1871*, P.P. 1871, LIX, Sect. III: Scotland in Civil Counties.
[2] FB Reps, *1854*, 34; *1880*, 20.   [3] FB Rep., *1886*, App. D, Table III.
[4] 1891 Census, Enumerators' Books.

sively of fisher origin. There was an evident tendency for the daughters of fishermen to take to other occupations, even among those who still remained at home; the boys, however, would still generally follow their fathers to sea.

Since the days of the first growth of herring fishing, one fishing area of the east coast had been distinctive and different—Caithness. The fishing population there grew quickly from 1780 onwards, partly by the concentration of landless men in Wick and Pulteneytown, partly in smaller groups in the secondary centres such as Lybster and Helmsdale, and even in still smaller settlements. Such men depended mainly on the summer herring fishing of the area. But from the forties onwards they might also participate in the early summer fishing of the west coast; from 1860 in the local winter herring fishing; and, throughout, in the small-scale haddock and cod fishings. A large part of the summer fleet, however, was owned and manned by crofter-fishermen who occupied the sub-divided cliff-top farms above the small coves and villages where the landless fishermen lived. Their holdings were small, amounting usually to no more than twenty acres on which two cows and a horse might be kept, and the rents were heavy, determined on the assumption that they could be paid 'out of the sea'.[1] Sometimes the farmer-fishermen would be sub-tenants compelled by the terms of their tenure to sell their produce to the tacksman. It was usual to fit out only for the summer herring fishing, the boats being hauled up on the beach for the greater part of the year.[2]

The crofter-fishermen held on through the first three-quarters of the nineteenth century, though often much in debt and only nominally owning the boats they operated. As long as the boats could be at least maintained in the local creek, and as long as they were cheap enough to form an investment which would be profitable solely through the returns of the summer fishing, then a regular and reasonably well-equipped fishing could be satisfactorily combined with small-scale farming (conducted on an individual basis rather than in the township frame that was common on the west coast). In the mid-eighties, therefore, these crofter-fishermen could still be described as typical.

The main obvious sign of the growth of the fishing population was the growing scale of the herring fishery, towards which the fishermen were being continuously drawn. For one thing the prices that were offered to fishermen by curers rose steeply from the mid-fifties, whereas

---

[1] *Commission to Inquire into the Condition of the Crofters and Cottars in the Highlands of Scotland*, P.P. 1884, XXXII–XXXVI, QQ. 37340, 37371–2, 37993.
[2] See figure, p. 82.

in the first half of the century rarely more than 10s. per cran was given; later a new minimum level of about 20s. per cran was reached, with annual fluctuations that might reach up to 26s.[1] Direct aid might also be granted by curers for the acquisition of new and larger boats and better gear.

Wick

Average catch per boat at the Caithness summer herring fishing

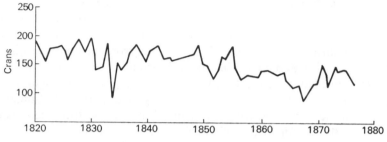

Source: *Daily Free Press*, 8 September 1890.

The first consequence was the continuing rise in the number of boats available for herring fishing, not only because there were more fishermen struggling to take part in the industry but also because the number of outsiders, hired simply for the summer fishing, increased even more than did the number of full-time fishermen.[2] In the late 1830s the herring fleet was evidently growing, but the rise was halted by the depression which started with the potato famine of 1845–6 so that by 1855 the number of first-class boats—the type that would be used for herring fishing—was, at 2,743, not very much greater than it had been in the 1820s.[3] But when prices began to rise from the mid-fifties nearly every one of the next thirty years was to see additions to the herring fleet. By 1886 there were 4,117 boats.[4] The process can be seen in detail in the figures for the Fraserburgh district. From 1835 to 1846 the number of herring boats increased from 130 to 139; there followed a decline but the rise had resumed by 1855 at the latest; and between then and 1877 the fleet grew from 197 to 328.[5] In the first period there was an unfailing annual addition to the number of Fraserburgh boats, and in the second in only two out of twenty-three years was there any falling back.

[1] *Daily Free Press*, 8 Sept. 1890.
[2] FBR, AF 24/119–125, Fraserburgh, Statistic Notebooks.
[3] FB Rep., *1856*, 24–39.   [4] FB Rep., *1886*, App. D, Table VII.
[5] FBR, AF 24/119–123, Fraserburgh, Statistic Notebooks. See figure, p. 83.

Fraserburgh

Annual increase in number of herring boats

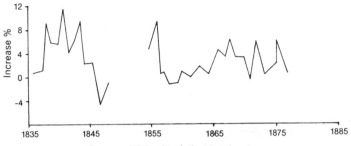

Source: FBR, Statistic Notebooks.

It was not sufficient, however, merely to increase the number of boats; continuously they were being built larger and equipped better than in the past, not only because this seemed to be one way to extract the last ounce of benefit from the situation but because natural conditions were changing so that in time better equipment was needed to secure the same effects. Early in the days of herring fishing it had been realized that bigger boats and more nets would pay off in bigger catches (or at least in staying the decline of catch) and slowly, through the decades, sizes edged upwards.

About 1840 boats were of a maximum of 35-foot keel compared with the 25-foot type of the beginning of the century, and they might carry thirty nets where twenty-four earlier had been the limit. By the early 1870s boats of more than 40-foot keel were being built. The big change of the sixties, however, was the adoption of cotton nets, twice as many of which might be carried in a boat of given capacity. In the Fraserburgh district, then, the average number of nets per boat remained constant, at thirty, between 1855 and 1860; by 1865 it had edged up to thirty-five, but by 1875 it had soared to fifty-five.[1] By the 1880s boats might well carry sixty nets and there is record of up to eighty—increases achieved partly by the substitution of cotton for hemp nets and partly by the continuing increase in the size of boats. A decisive change—increasing freedom of operation—came in the seventies with the adoption of decked vessels which, if they were of the same dimensions, would carry less in nets and in catch than the undecked. With their adoption the pressure to enlarge the size of boats became even stronger. By the seventies the 45-foot boat was common and by the 1880s the new Zulu

[1] Ibid.

and Fifie types were commonly of 55-foot keel length and, exceptionally, might be of over 60-foot.

The use of large-scale and expensive equipment was connected with another series of changes which widely affected the lives of fishermen— the tendency of the herring fishing to spread into new areas and to occupy more and more of the fisherman's year. On the one hand, the simple and unavoidable response to the desire to increase the size of the operating fleet was to find new areas of fishing; on the other, expensive and specific equipment could only be made to pay its way when used over a considerable part of the year.

Fraserburgh

Ratio of catch per boat to value per boat (boat and gear)

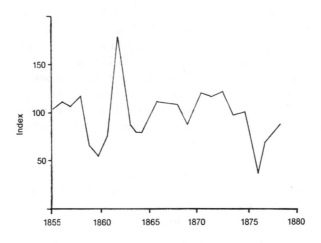

Source: FB, Annual Reports.

Thus the shoals that collected at many points along the east coast during the two-month period beginning in late July were attacked on a broadening front. By the early 1820s the first expansion of the fishing area had been brought to an end when for a number of years the great southerly swing out of the early Caithness enclave stopped short. The main area of summer fishing then stretched from northern Caithness to just south of Peterhead, while at the other extremity of the coast, in Berwickshire and East Lothian, a long-standing summer fishing continued to give an output small in relation to the rising flood farther north. In the twenties a winter fishing for herring was established in

Fife but again the scale was small and until the mid-thirties there was no further growth in the area of herring fishing.[1]

Then, about 1835, the fishing fleet, drawn from every part of the coast, began to lap into new areas. Stonehaven, south of Aberdeen, attracted a few nearby boats for a herring fishing; but far more important was the emergence of Fife as a main area of summer operations, with the curing of the product of a local fleet that had hitherto been among the most active and mobile in searching out good fishing areas to the farthest northerly point on the coast.[2] When Montrose in 1861 and Aberdeen in 1871 rose to be large herring fishing centres[3] no extensive section of the east coast then lacked local stations where curers offered a market for herring during the period of summer fishing, from late July to mid-September.

Yet the pattern never became stereotyped and was never so shaped that local effort exactly matched local resources of men and boats. The fleet was made up of many tiny constituent parts drawn from over a hundred distinct and separate communities, each providing crews ready to move from one region to another, with every year seeing a complex movement from one part of the coast to another. These migrations were never in quite the same pattern from one year to the next and the balance of the forces collected at the different centres was always shifting. Thus the upper section of the Moray Firth from Macduff westwards, which had been an important and attracting area from the early days, was progressively deserted by the main fleets from the mid-fifties. Fife's period as a major centre for summer fishing was relatively short, with decline starting in the mid-sixties. Caithness declined in relative standing although it never ceased to be an important area. The tendency of the seventies, after establishment of bases all along the coast, was for activity to concentrate on Caithness, on the north-east corner from Macduff to Montrose and on Berwickshire.

II

In the early forties, a move of another type started when the east coast boats began to cross in force to the fishing grounds lying to the west of the country and initiated a fishing that was additional to the activity of the summer months in home waters.[4] The main fishing area

[1] *Fife News*, 22 Jan. 1872.   [2] *NSA*, IX, Fife, 979.
[3] *Report on the Herring Fisheries of Scotland by the Inspector of Salmon Fisheries for England and Wales, and the Commissioners of Scotch Salmon Fisheries*, P.P., 1878, XXI, p. xii; Edmund W. H. Holdsworth, *The Sea Fisheries of Great Britain and Ireland* (London, 1883), 162.
[4] *John O'Groat Journal*, 24 May 1844.

was the Minch and the bases first chosen were in Lewis. The season began in May and continued through most of June so that the period of the year covered was quite different from that of the established east coast season in which the crews were also invariably involved. When the east coast men ventured thus out of their home waters they had to adapt a method of fishing and an organization tried on their own coast through nearly fifty years to the different conditions of the west coast. The result was a system somewhat different from that of the east coast but also entirely novel to the west coast. The fishing grounds that were first exploited were in the Minch, farther to sea than the traditional areas of herring fishing on the west coast, but well fitted to the standard east coast practice of making daily landings at shore stations fully equipped for all the business of curing. Stornoway, with an already developed harbour was used as the first main base but before long the curers were turning their attention to more isolated and empty spots north and south along the east coast of Lewis.[1] In such cases the whole apparatus of curing, together with dwellings for workers and for crews, had to be temporarily fabricated. The facilities of even sheltered lochs were greatly improved for herring fishing by the erection of simple jetties at which unloading could be carried out from deep and sheltered water. At the cost of about £200 a curer could create such an apparatus solely for his own business. Curing stock, in the shape of salt and empty barrels, was transported from the east and landed in a quantity estimated to be sufficient for the needs of the season. Provision had also to be made for living quarters both for curing staff and for crews. Coopers would come from the east while female gutters came from the islands; even the latter, however, might come from a considerable distance and had to be housed, and to this end pre-fabricated huts would be carried from the east coast base to be erected on the empty shores of the loch.[2]

The west coast fishing expanded from its first base in Stornoway into the more remote lochs of the eastern side of Lewis, so that by 1860 twelve bases had been developed for use over the six weeks of the early summer season.[3] Then a distinctly different cluster of stations was developed in Barra.[4] Again it was the shores of empty but sheltered sealochs that provided the ground for curing and the whole working station was temporarily fabricated by the landing of stock and materials brought from the east. The season was the same as in Lewis and the

[1] *Report on the Herring Fisheries, 1878*, p. xviii. Also see map p. 102.
[2] R. J. Duthie, *The Art of Fishcuring* (Aberdeen, 1911), 23–30.
[3] *Report on the Herring Fisheries, 1878*, p. xviii.
[4] Ibid.

Barra fishing was, in fact, alternative rather than additional to those previously established for the summer season. The Barra and Lewis fishings at their peak employed some 1,400 boats, four-fifths of which were of east coast origin.[1] Altogether, about one-third of herring boats owned by east coast men might be taken to the west coast in the early summer.

Again, in the sixties east coast crews also began to move southwards to make trial of the English fishings, particularly in the autumn. The Yorkshire stations attracted some Scottish boats even at a season when their home stations were still active, but the most important move came when Fife boats journeyed to East Anglia to share in the fishing of that major area.[2] For the Fife fishermen this became an important annual venture and before the sixties were out over 100 boats, carrying crews that included most of the active fishermen of the region, would make the long journey south.[3] But the attractions of East Anglia made little impact on fishermen from other parts of the coast till nearly the end of the century.

In the early eighties a new area of prolific yield opened up to the east coast fishermen—Shetland. When in the late 1870s the tentative trials in these waters showed high potential for fishing conducted at almost any time between May and September, there followed a rush to develop new stations and a build-up in fishing strength quicker than any previously known.[4] By 1884 a fleet of 900 boats was operating off the islands. The speed of the expansion in Shetland was due partly to the flexibility which the east coast crews and curers had been learning since mid-century, partly to the fact that the coastal configuration allowed the quick assembly of efficient stations on the model that had been first tried in Lewis, partly to the great size of the catches and partly to the ready response by the fishermen of the islands themselves. The Shetland herring fishing season was divided into two periods, continuous with each other but distinct in organization and in the herring stock on which they depended. The early fishing of May and June mainly centred on the northern islands; then came the switch south to Lerwick and the more southerly stations for the period of summer fishing starting in July. A large fleet of boats from the Scottish mainland would be engaged by curers to go to the early summer fishing in the north; some of these would return for the home fishing in July, but some would remain in

[1] FB Rep., *1888*, 21.
[2] FBR, AF 19/23-24, Weekly Reports, Anstruther, 9 Sept. 1865; 15 Sept. 1866; 8 Aug. 1868; 30 Oct. 1868.
[3] *Fife News*, 4 Oct. 1873; 25 Sept. 1880.   [4] See below, pp. 201-4.

Shetland until September. By the early nineties the older system of engagements by curers had given way to daily auctions in Lerwick. But in the outer voes engagements continued to be the dominating principle for much longer.[1] With the growing sensitivity to opportunities of fishing in the different and competing areas, the numbers going to Shetland (as to any other particular fishing) fluctuated from year to year.

Another extension of the herring fishing season was achieved by the development of winter fishing. Fife had been the first area to establish such a fishing in the 1820s, but in the 1860s—starting with Caithness—all the main regions began to operate a local winter fishing on a modest scale.[2] Normally there was little migration to participate in this fishing and a smallish proportion of the local men would form crews, entirely composed of full-time fishermen, all with some investment in the boats or the nets in the venture. They then operated from the local ports. In this fishing there was no specified engagement to curers and landed fish were sold on the quay. The outlets for the catch were much more diversified than with the summer fishing, a fair proportion being sold as fresh fish—even from the most northerly stations which had now been reached by the railway, some being reddened, a little cured and some sent south, sprinkled in salt, to be converted into bloaters.[3]

III

For all the impetus given by herring fishing, the older types of fishing, with small- and great-lines, continued to occupy a place, and sometimes a dominant place, in the calendar of every east coast fisherman. The pursuit of haddock and cod in particular occupied much time.

By the 1830s the processes and the markets which made haddock fishing worthwhile had merged in a pattern which was to be broadly sustained till the coming of steam-trawling in the eighties. The smoking of haddocks by methods which were first developed in the cottages of the North East had spread along the more southerly portions of the coast, and particularly into Fife and the Eyemouth district, although there it was always the independent curer who undertook the smoking. The main market for these southern factories, turning out their own version of the 'Finnan', was in the industrial west, the cured fish at first being carried across the country by cart. By the forties the railway was replacing the cart but this tended merely to strengthen the existing

[1] FB Rep., *1894*, 172; *Daily Free Press*, 23 Dec. 1887.
[2] *Fife News*, 27 Jan. 1872.
[3] *Sea Fisheries Commission, 1866*, QQ. 29108, 28857, 30177–80.

links.[1] In particular, Eyemouth flourished on the basis of the regular daily service that was provided by the railway[2]: fish which had been landed there in the afternoon would be processed with the light two hour smoking required for the Glasgow market and transported by rail to be sold by the retailer on the following morning.

In the North East, where originally the haddocks had been smoked within the cottages themselves, the curer was already by the thirties taking some of the catch to be processed in his own factories; and he seems to have been no more careful than were the curers in the south to match precisely the finer product of the cottage industry. Yet in this region some of the catch continued to be treated under the older system in which each fisherman cured his own share of the catch. The resulting product was in part sold locally by fishwives doing their own retailing, but some would be transferred to grocers who sold to a local clientèle, some went south to England in the steamships and some began to be taken to the south-west of Scotland. It was 1850, and in some cases as late as 1890, before the fishing centres of this region had direct connection by rail with the south and the tendency then was to strengthen the pull of the markets of the industrial west of Scotland.[3] Yet the pattern of process and market remained solidly fixed till the eighties when, for example, some of the haddock were still being smoked in the cottage 'lum' or kiln to be carried to Aberdeen for sale to grocers, while some of the crews would have contracts with curers for the sale of the whole product of six months' fishing at pre-arranged prices.[4]

Within this framework, prices tended to be lifted by the improvement of transport to, and quicker communication with, an industrial population which was growing rapidly. No complete price series is available for haddock but when any change shows in the occasional recordings it is invariably an upwards move. Thus in the 1840s the normal price for haddock was cited as 4s. per cwt.[5] By the mid-sixties, 8s. was being given and prices remained near that level[6] till the early eighties when further sudden rises took price, at least on occasion, to 12s. per cwt.[7]

[1] *Fife News*, 12 Oct. 1872.
[2] *Sea Fisheries Commission, 1866*, QQ. 28036, 29069; *R.C. on Trawling, 1884-5*, App. B. pp. 435–46; QQ. 6987–93; *Fish Trades Gazette*, 12 Oct. 1872.
[3] *Sea Fisheries Commission, 1866*, QQ. 28073–85; *Fish Trades Gazette*, 7 July 1883.
[4] *R.C. on Trawling, 1884-5*, App. B 1, p. 440.
[5] Thomson, *Value of the Scottish Fisheries*, 65–6; *R.C. on Trawling, 1884 5*, QQ. 1968, 1983, 2023, 2046, 2221.
[6] *Sea Fisheries Commission, 1866*, Q. 24791.
[7] *R.C. on Trawling, 1884-5*, p. 939; *S.C. on Sea Fisheries*, P.P. 1893–4, XV, Q. 7655.

In the North East the fishing for, as well as the processing of, haddock tended to remain traditional. Much of the catch was made in small boats, operated directly from the home villages and taken daily only a short distance to sea. Each village in fact, whether or not it had a harbour or a pier, would have its local haddock fishing in the winter months. A primitive type of fishing economy tended to persist in which boats had to be hauled over the beach and in which, necessarily, the womenfolk were engaged in the baiting of lines as well as in helping with the launching of boats.

Auchmithie gives an example of the old-fashioned economy still working in the eighties.[1] The boats would set off from the open beach at 2 a.m., the men being carried to the boats on the backs of the women; by 10 a.m. the crews would be back with the daily haul. The one sign of modernity in this little centre was the use of a steam engine and hawser to haul up the boats on the beach. There followed intense activity by the whole family around each fisherman's cottage. One woman would decapitate and gut the haddocks, another would slit and clean them, another put them in salt in a tub and yet another tied them by the tails in pairs on a pole. The fish were then hung on a scaffolding to dry before being sent for sale to Dundee. Part of the catch would be smoked for an hour on wooden rods stretched across small wood fires burning in excavated holes or ovens near the owner's cottage. The fish were then packed in tubs also to be taken to Dundee. The baiting of the hooks, of which there were 6,500 to each set of lines, was done by the children.

In Fife, and more particularly in Eyemouth, larger boats were used for the haddock fishing. At Eyemouth, indeed, this became the occupation of nine months of the year. Some of the largest and best-equipped boats in Scotland were taken daily to the deep water and the catches with which they returned provided good and steady incomes.[2] But although in Fife and farther south the haddock fishing might be centralized because of its growing dependence on harbours that could accommodate the large boats, it still had to rely on the labour of the fisherman's family for the baiting of lines. Indeed, the great increase in the length of line and the number of hooks made such help all the more necessary. In Eyemouth labour could be hired for the purpose but it was expensive and, where possible, the fishermen made use of their own, unpaid families.[3]

[1] *Fish Trades Gazette*, 22 Sept. 1888.
[2] *Berwick News*, 8 Feb. 1870; 20 Dec. 1870; 2 Mar. 1880; *Sea Fisheries Commission, 1866*, QQ. 28073–85; *Fish Trades Gazette*, 7 July 1883.
[3] *R.C. on Trawling, 1884–5*, Q. 6966; *S.C. on Sea Fisheries, 1893–4*, Q. 6358–9.

By mid-century, great-lining (with cod as the main objective) was tending to decline, but by 1870 there were signs of some revival in what had once been the main source of income for east coast fishermen. Cod and the more valuable great fish such as halibut gained considerably from quick transport to the south; and the 1870s, for example, saw a big jump in the consignments of fresh cod going south from the Fraserburgh district (the railway to the Fraserburgh terminal having been opened in 1865).[1] Even the curing of cod and ling by drying seems to have increased when the price of dried fish rose in the same decade. Drying was now carried on increasingly, but not entirely, by curers and the process was in some cases speeded up by the use of racks rather than the open beach, although the older method of 'steepling'— alternating with the laying of the fish in extended order—was still being described as typical in the eighties. And in Anstruther in 1872 we hear of every wall being covered by drying fish.[2]

The catching of cod, unhampered by the need to have lines baited ashore and pursued most productively at the relatively distant banks, tended to be undertaken in the larger type of boat and therefore to be centralized in the main ports. When the Union Harbour of Anstruther was opened in 1877 it became the main centre of the great-line operations of the Fife men and, indeed, this was a main source of income for the fishermen of the region.[3] The harbours to the north such as Montrose, Peterhead, Fraserburgh and Buckie each served as a base for the great-line activities of the fishermen of their hinterlands, but cod fishing continued to be pursued by a relatively small proportion of the crews of this region. Haddock fishing at the small creeks occupied possibly a greater number, even in the spring months when the cod fishing was at its height.[4]

Fishermen had a growing power of choice, then, in the type of fishing and increasingly they would adapt their efforts in a sensitive appreciation of a changing complex of opportunities. Herring fishing was the almost invariable occupation of the late summer months, although every year the crew had to make a choice of the station where they wished to fish. The curers played a part in determining the outcome of this choice by the terms they offered, but they had no power of direction. The crews were left to weigh one thing with another, and in the end would agree to one bargain among several offered as to place of work

[1] FBR, AF 24/95–114, Fraserburgh, Private Books.
[2] *Fife News,* 15 June 1872.   [3] *Fife News,* 25 Mar. 1871; 1 Apr. 1871.
[4] *R.C. on Trawling, 1884–5,* QQ. 929–35; *Demand for Scotch Cured Fish,* P.P. 1888, XCVIII, Q. 395; *S.C. on Sea Fisheries, 1893–4,* QQ. 929–35, 6355–9, 6490; *Fish Trades Gazette,* 30 June 1883.

and as to employer. In autumn many took to the haddock fishing, which was very localized and might last for a full six months, but from the sixties there was also the possibility of trying the East Anglian venture till Christmas. In January and February, for those not committed to a continuing haddock fishing, there was the winter herring fishing; crews never went far from home to participate in this fishing but from the sixties most parts of the east coast had a small herring fleet working from the main ports. Then in March came the great-line fishing, widely spread but conducted in the bigger boats and therefore from the main ports; again the main competition was from the haddock although the great-line season might overlap with the early summer fishing for herring on the west coast and, latterly, in Shetland.

If it was left to the individual crew to settle its own routine, nevertheless certain definite regional preferences emerged. In the North East there was great emphasis on the haddock and many would occupy the whole period from October to April in its pursuit, using small boats which in the eighties might still be worked from open beaches. It remained a very dispersed effort, conducted from the home village and still making great calls upon the labour of the whole family.[1] It was from this area, too, that boats were mainly drawn to the west coast and the Shetland herring fishings. The commonest routine, perhaps, was summer herring, winter haddock and early summer herring in distant waters, although the area had also a certain amount of great-line fishing. In Fife, the autumn voyage to East Anglia was popular, winter herring was a pursuit from early in the century and great-line fishing increased in the late sixties and seventies. The haddock was here sought in big boats working far offshore but on the whole it played less part than farther north. To the south, Eyemouth had its own very particular routine in which large boats fished for haddock for nine months of the year, the remaining time being taken up by the local summer herring fishing.[2] Farther west, along the coast of the Lothians, herring fishing played a big part with the crews making long migrations at several times of the year.[3] Caithness was in the somewhat peculiar position of retaining a body of fishermen who were also farmers and who might fit out only for the summer venture; but the area supplied boats manned by its full-time crews for the west coast and for Shetland, together with the local winter herring in January and February.

[1] *R.C. on Trawling, 1884–5*, Q. 954; *S.C. on Sea Fisheries, 1893–4*, QQ. 6355–9, 6490.
[2] *Sea Fisheries Commission, 1866*, QQ. 28036, 29069.
[3] *R C. on Trawling, 1884–5*, QQ. 4772, 6586.

IV

The incomes of fishermen during this period are very difficult to calculate. A crew might well vary the seasonal pattern of its fishing from year to year, while in a given year particular crews would follow different sequences of fishings. The single common factor was the summer herring fishing and for this invariably nearly every first-class boat would be prepared. It was this fishing, too, which dominated calculations about required investment and expected income.

The equipment for herring fishing was being made continuously (and sometimes quickly) more efficient; yet there was no clear response in

Fraserburgh

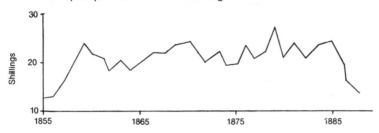

Fraserburgh

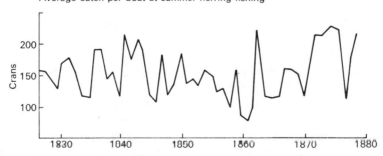

Sources: *Sea Fisheries Commission*, 1866, App. p. 699; FBR, Weekly Reports: *John O'Groat Journal.*

Fraserburgh

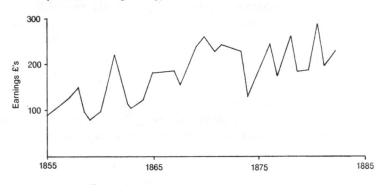

Average gross earnings per boat
(summer herring fishing)

Fraserburgh

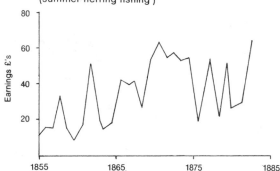

Estimated income per (owning) fisherman
(summer herring fishing)

Sources: FBR, Weekly Reports, AF 37/149.

increased catch. Indeed until about 1870 the trend of catches at the summer fishing was very slightly downward.[1] The increasing difficulty of catching herring from old-established stations was also evident in the fact that boats were being forced to fish at considerably greater distance from shore; offings of more than thirty miles became the common nightly resort of the boats. Until the late sixties, then, the fisherman was having to work harder and to spend more to bring back the same quantity of herring. Then in the seventies came the apparent reward of the pouring of resources into the improvement of the fleet. With occasional disastrously bad years, catches moved to generally higher levels.

[1] See figure, p. 93.

Part of this gain was cancelled out by the fact that the boats now were each carrying one more man in the crew, but there can be little doubt that the yield per man of those engaged in summer fishing increased perceptibly in the 1870s and 1880s.

Although the output of each man for the summer's work moved up very slowly and intermittently and only by virtue of heavy investment, a basis for rising incomes was to be found in the trend in the price of cured herrings which, as we have seen, generally although not continuously raised the price per cran paid to the fishermen. Thus, at least from mid-century, the gross earnings from the summer fishing tended to rise. The fluctuations, it is true, were unceasing and large; and a poor year late in the century might well be worse than even a moderate year in the fifties. Nevertheless, on a calculation based on average catches and on prices paid to fishermen, in the eleven years up to 1855 on only two occasions did estimated individual incomes rise above £30, while in the eighteen succeeding years on only four occasions did it fall below £30.[1]

The earnings of owner-fishermen were made uncertain and on the whole were brought under increasing pressure by the haphazard movement and the long-term upward thrust in the wages that had to be paid to hired men. The number of potential labourers gathered in the ports might be excessive or deficient to the needs of the boats collected there in somewhat unpredictable numbers. For example, in 1853 wages jumped by 25 to 40 per cent because of a shortage of hired men amounting to about 200.[2] There were also big differences at any one time between the remunerations of particular individuals. Thus in 1854 in Wick 'prime' hands would get £7 to £8, secondary £5 to £6. 10s., and third-rate £3 to £4. 15s.[3]

Yet some long-run changes can be discerned. In the 1840s £6 was spoken of as a top wage, whereas by the mid-fifties this had increased to £8[4] which was still considered to be good remuneration until the mid-seventies. Then, while the rates were normally no better than they had been for twenty years, earnings might on occasion be much higher. This might be because the basic rate was raised to levels of £10 or £11, but more commonly it arose because of the tendency to pay more of the wage as a rate on each cran handled; sometimes the whole wage was determined in this way and the rate might then rise as high as 2s. or even 2s. 6d. But, more usually, a smaller cran rate (of say 1s.) was

[1] See figures, pp. 93–4.   [2] N. Ensign, 21 July 1883.   [3] N. Ensign, 20 July 1854.
[4] John O'Groat Journal, 21 July 1848; FBR, AF 36/23, Wick Weekly Reports, 20 July 1867.

added to the basic seasonal payment. With the high catches of the late seventies and early eighties the seasonal remuneration might reach the unprecedented level of £30.

Fraserburgh

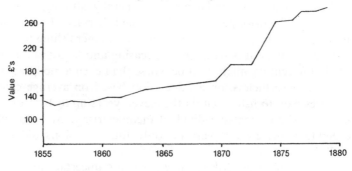

Value of boats and gear per owning fisherman

East Coast

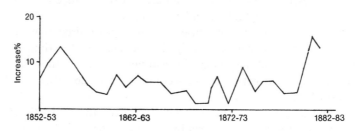

Annual increase in total value of boats and gear

Source: FB, Annual Reports.

The summer herring fishing represented the earnings of rather less than three months of the year. To his earnings in that period any fisherman would have to add the proceeds of two or three other fishings —herring fishing on other coasts, winter herring fishing, haddock or great-line fishing. Until the Shetland boom, which only began to affect a large number of mainland fishermen in 1882, the outer herring fishings would generally give a lower average return than the summer fishing at home; individual earnings at the Minch fishing of early summer would seldom be more than £25. And the haddock fishing seems to have given, rather more regularly than herring fishing, a return of between 10*s*. and 15*s*. per week.[1] Thus even in the seventies, from the base of £50 for the summer fishing, a man's earnings would be unlikely to top £100 for the whole year; and out of this had to come the hefty charges for the upkeep of his equipment.

Fraserburgh

Index of equipment per head (real terms)

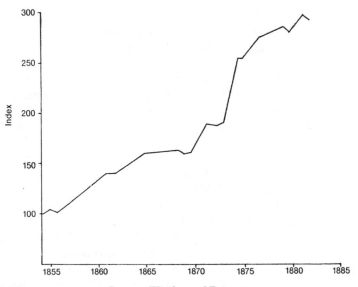

Source: FB, Annual Reports

While incomes were thus rising slowly and uncertainly the capita costs which faced fishermen mounted. The improvement of boats and gear meant a corresponding, and sometimes more than proportionate, increase in expense. Whereas a first-class boat fit for herring fishing could be purchased for less than £100 in the late forties and fifteen years later it was not usual to go beyond £150 for even the best boat of the day, by 1880 a first-class boat would cost £300 and the eighties were to see further sharp increases—particularly when the steam capstans were added. Similarly, over the thirty-year period from 1850 to 1880 the drift of nets had more than doubled in size and consequently in cost.

[1] For example, *R.C. on Trawling, 1884–5*, App. 3, 1, p. 240.

It should be remembered, too, that at least until 1870 the costs of each boat were tending to be shared by a smaller number of partners. Thus, on the figures given by the Fishery Officers of the Fraserburgh district, the average herring boat and its gear as held by the fishermen was worth £132 in 1854; by 1860 the figure was £162; and by 1878 it was £260.[1] For the east coast as a whole, the value of equipment increased by 90 per cent between 1854 and 1869, and by 97 per cent between 1870 and 1882. Per head the increases were slightly less—at 43 per cent and 62 per cent for the same two periods—but were still striking. All told, in the thirty years the average value of equipment being used, mainly for herring fishing, by the east coast fishermen increased by 144 per cent.[2]

Such increases were met mainly by the individual saving effort of the working fisherman, who thus secured his independence at a time when large-scale capital was becoming so much more important. The underlying process of finance depended on the sharing of costs as well as profits among the general body of the fishermen. It continued to be a settled principle that boats and gear were provided by the fishermen themselves and that all full-time fishermen—not including the hired men who came from outside the fishing community—contributed to the total stock in roughly equal proportions. On the ordinary fishermen therefore devolved the duty of providing the increasing increments of capital needed for efficient fishing. The pattern in which shares in the boats were divided varied somewhat from one region to another. In the North East (including the Moray Firth) it was customary for boats to be owned by two, three or four partners who were in fact often related by blood; in Fife one man might well be the sole owner, although he was always a fisherman and other members of the crew would provide gear; in Eyemouth the whole crew of seven would be equal sharers in the haddock boats.[3] Only in Caithness did outsiders, in the shape of curers, own boats in part or in whole.[4]

The provision of nets was a separate problem. Where boats were owned by all the full-time fishermen in the crew, the nets of course would be provided by the same men. But nets might be a man's sole investment in a fishing venture, either because of the separation of owners from the others of the crew or because a venture might be arranged in which men who in fact owned a boat would join in another

---

[1] FBR AF 24/119–123, Fraserburgh, Statistic Notebooks.
[2] FB Reps. See figures, pp. 96–7.
[3] *Second Report of the Commissioners appointed to Inquire into the Truck System* (*Shetland*), P.P. 1872, XXXV, Q. 16701; *Fife News,* 19 Apr. 1873; *Fife Herald,* 24 Aug. 1871.
[4] *Commission on the Truck System, Second Report, 1872,* 295–7.

boat by simply putting in the nets. In any case nets, being as expensive to maintain as boats, were the basis of independent investment and share. Since fishermen might well use more than one class of boat in the year they were also quite likely to have shares in more than one. Given, then, that nets and boat represented different investments and that one man might have shares in two or even three boats, the whole pattern of sharing becomes immensely complicated as well as variable from one region to another. But the complications do not invalidate the basic and traditional principle that all fishermen had shares and thereby provided all necessary boats and gear.

Fishermen, of course, were able to call on help in finance. They received no direct loans from outside the industry, but they depended heavily upon curers for advances for the first acquisition of boats. Nets, on the other hand, were frequently paid for out of the bounties which were given on engagement. The curers in fact supplied some sort of link between the needy fisherman and the outside financial world, for even while they were making loans to the fishermen they were also drawing upon credit from certain restricted sources outside the local confines of the industry. As we have seen, short-term loans came from German importers and from banks. Although such money was mainly used as a seasonal support in the curing process, indirectly it might help in the process of acquiring fixed capital in the shape of boats and nets; curers with their requirements for short-term funds met by borrowing were better able to make the longer-term loans to fishermen by which boats were purchased. Yet while fishermen did often seek help in the acquisition of boats—and, as we have seen, they acquired boats frequently and at increasing expense on each occasion—they did not thereby cease to be full and effective owners. It was in interest-bearing loans that they received the money; and the curers as lenders, while obtaining first right to the catch for the fishing season, did not directly share in the proceeds. The obligation of the crew to fish for the curer from whom they had borrowed was nevertheless an important restriction. Such a crew lost the power of moving about and making the best possible bargain in an open market and were generally paid a lower rate for landings than were the free crews. The evidence, however, is that in all districts other than Caithness repayments were made speedily. Effectively, apart from temporary aid, the fishermen were providing the means by which the great accumulation of expensive equipment was acquired. In so doing they retained their independence as share fishermen, moving freely, making the best bargain they could with the curers, and dividing among themselves the full proceeds of the various fishings.

They could become, too, people of considerable property: a man might have shares in two or even three boats, together with an appreciable proportion of the drifts of nets and of both small- and great-lines which he might use at different times of the year. Thus by 1884:

Well-to-do fishermen are sure to possess, first, a house and furniture; second boat and gear, or perhaps shares of a large and a small boat; third nets, lines and other fishing materials. The heads of families are generally tolerably comfortable as regards means. A small proportion may, through unfortunate circumstances, be poor for a time, but perseverance soon overcomes the poverty. The various banks receive a goodly amount of money on deposit from them; and when we consider that mostly all the houses in the fishing villages which they inhabit belong to themselves for the greater part, we must allow that as a class they are both powerful and rich. Young fishermen, as they earn and save money, invest it in their friends' or relatives' boats, thereby securing an interest in the boat, and therefore in the industry.[1]

[1] Miln, 'Scotch East Coast Fishing', 6.

CHAPTER VI

Crofting and Fishing: the West Coast, 1790–1884

I

AROUND THE west coast as a whole two regions of rather distinct social features can be distinguished. One comprises the lochs and sounds into which the Firth of Clyde divides—a region where the influence of the Glasgow market has generally been paramount. The second is the much more sprawling area of mainland coast and islands which lies west and north of the Kintyre peninsula. Though there are considerable differences between its sections, the latter area has common features which give it a unity in contrast to the Clyde region. The main fishing of the area has generally been carried on north of the island of Mull and for convenience we will call it the 'North West'. It is to this region that we turn first.

One notable difference between the east coast and the North West was that the small-boat fishing, the typical activity of the coast, was in the hands of a people whose main interest was in the working of land: 'Their sole attention is in a manner fix'd to the produce of Earth, their sole object is to get a Farm, and a patch of ground however small is infinitely prefer'd to any other mode of gaining a living.'[1] This preoccupation was to continue to permeate all west coast fishing to the north of Kintyre and all plans had to take account of the feeling. Yet it was not a condition necessarily inimical to a growing industry. As a result, a large population was interested in fishing and lived where this interest could turn into effective action.

Much of the land of the western Highlands and Islands is, of course, high in elevation or rough in surface and only a tiny proportion is cultivable. The arable stretches—and the notion of a farm always included an arable section which could be used to grow subsistence food crops—mostly border the sinuous coasts, either in isolated patches or in more continuous stretches. Such land, however small the contiguous cultivable area, was always arranged in group settlements

[1] BFS, Vol. 3/90, 'Extracts from letters', 1787.

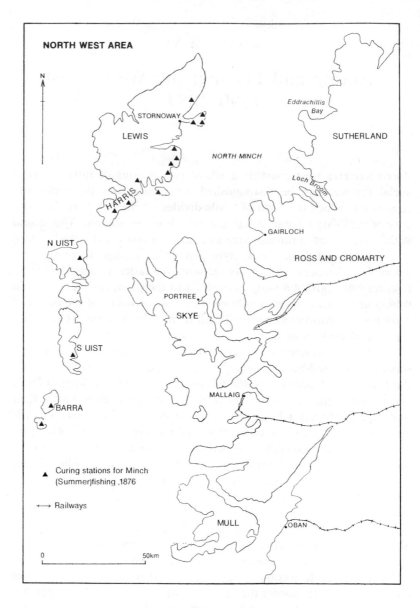

Source: FBR, AF 37/153.

rather than in isolated farms. The typical settlement of the eighteenth century consisted of a huddle of cottages on the seaward side of an irregular patchwork of fragments of cultivated ground. Sometimes such a settlement would sit many miles distant from its nearest neighbour, alone perhaps at the head of a sea-loch; sometimes it would lie close to others of its kind. But in any case the individual family lived as one of a cluster of families with which it would associate in the activities that demanded the labour of the group; fishing crews were easily and naturally raised from among the clustering small-holders. Most of the cottages had easy access to the water and boats were needed for purposes of transportation as well as for fishing.

When agricultural improvement came to the region the settlements were re-shaped. From the 1770s onwards landlords in one area of the Highlands after another turned their thoughts to improvement and had the land laid out in lots, essentially separate from each other but of such a size and shape that each had an edge to the shore. If the cottages of the holders were set in an extended line rather than a cluster, they were still close enough to each other to form a relatively compact settlement in which the inhabitants would still find it easy to associate in common tasks and in which all had equal and easy access to the water. Another factor, becoming evident in the later eighteenth century and continuing in the North West till at least 1850, was increase of population. Since the rage for sheep farming reached the area about 1800 and brought many clearances from interior farms, the increased population was subjected to conditions of ever more intense crowding in the coastal settlements. Sometimes, as in Harris and Sutherland, new holdings were laid out for the evicted—lots which were deliberately fashioned to provide bases for fishing, being tiny in size and often inferior in quality of land. Even though the people were crowded on the coasts and holdings were divided into smaller and smaller fragments, and even though an increasing proportion of all families were forced to subsist without any recognized holdings, the primary interest in farming never relaxed; most families obtained the use of some land and, however tiny the fragment, it remained the centre of their economic life. Yet circumstances also forced a continuing and even an increased dependence on fishing, and the very fragmentation of holdings intensified the need to find other sources of income. Subsistence was never secure and the ultimate disaster of famine in the 1830s and 1840s merely emphasized long-standing difficulties. Rents, too, were increased in the period of high prices up to 1815, to levels which could not be met out of the proceeds of farming; every possible source of cash income had to be sought.

Fishing, however, remained secondary—an occasional and very often a highly chancy venture for the people of the North West. The large number of boats owned by the people makes a poor indicator of the depth of the involvement with fishing. Boats were used for many purposes and it might be only occasionally that they were turned to fishing. Normally the small craft, of 14- to 16-foot keel, would each be owned by a group of three or four men who would share as equals both the expenses and the profits of fishing.[1] In most townships the greatest preparation that would usually be made for fishing with these boats was to have ready not more than twelve nets which would serve as a small drift. Thus perhaps one family in three would have some share— say a third or a half—in a boat which might very occasionally be used in a herring fishing. Such fishing was worthwhile only because, when the herrings came, it was in great shoals which in sheltered waters gave good catches even with light gear and frail boats.

The pursuit of the herrings in this way was the main fishing interest of probably the bulk of the part-time fishers of the North West. It was a desperately uncertain enterprise, with each crew depending on the herring shoals entering its own particular loch. The winter was the usual time when the herrings might be found in the loch but the timing, or even the eventual fact, of their approach could never be certain. Quite unpredictably, particular lochs would swarm with herring for a few days or a few weeks or would be left blank and empty throughout the winter. Thus, with boats that were confined to working within a very small radius, the catch fluctuated from nothing at all to an abundance in which it was difficult to use or to dispose of all that was landed.

The first aim was to cure one or two barrels for the use of the household, but even this simple objective might be unattainable because of shortage of local supply or because of lack of salt. Almost inevitably the fisherman needed the help of the curer and when any large landings were made it was only by considerable preparations that they could be given any value. But curers also were frustrated by the perplexing variations in the local fishings. Pickling, which could be performed with light and transportable apparatus, seemed to be the form of curing best adapted to a region where the local occurrence of good fishings was always uncertain. Red-herring houses, such as were erected in Loch Broom, were rare and absorbed the catches of only a tiny body of fishermen.[2] Even pickling, as we have seen, depended on the availability

[1] OSA, III, 373; IV, 574; VI, 290; XIII, 559; XVI, 174.
[2] Committee on the British Fisheries, 1785, 141; BFS, Vol. 3/71, 'Extracts from letters in reply to printed queries', 1787.

of a bulky and fairly valuable stock of barrels and salt. No curer, then, could possibly make provision at all the possible points of fishing for the reception in full of the maximum landings that might there occur; and a large proportion of his stock would lie unused through the season. The best hope was to hold stock in readiness to be rushed to any site where good fishings were reported. Sloops could be laden with salt and barrels in preparation for being moved to where they were needed, and this was a manner of operation inevitably more costly than the fixing for the season of shore stations—the system that was shown to work well in the different conditions of the east coast.

In spite of the difficulties, preparation was made every year to try to cure as much as possible of the uncertain catches of the North West. Small vessels of up to 100 tons would be fitted out—some with nets as well as salt and barrels, some only with curing stock—mainly in the ports of the Clyde estuary. This was in addition to the 'busses', vessels of similar type which were fitted out and manned under the terms that qualified them for the tonnage bounty. The prime object of the sloop fleet was the curing rather than the catching of herring.

Every year, then, a considerable fleet was fitted out to move according as the herrings showed themselves, depending on the supplies provided by the boats of the particular locality. When reports came of good fishings, sections of the fleet would converge for action. Stores would be landed by the loch, 'yards' would be laid out, and labour would be engaged. In the meantime, boats might come in from the surrounding lochs. Often more than 100 boats would group for the fishing and in some cases fleets running into several hundred boats were collected.[1] The herrings would be cured either on the vessels or on shore, being purchased from the boats by the curers at a daily rate determined by the supply of the moment in relation to the curing facility. Finally, after days or weeks the fishing would be played out or perhaps the curing stock exhausted. The boats would disperse and the sloops head southwards with their cargoes of cured herring. Sometimes they would be full to capacity; more commonly they would have used only part of their stock and would bring back less than their full lading of barrels. In any case the profit of the season for the curer depended on how far he had used up his stock.

It is difficult to discern the progress of such a spasmodic fishing in terms either of its impact on the life of the local community or of its contribution to the national output but it is evident that no outstandingly

[1] James Wilson, *A Voyage round the Coasts of Scotland and the Isles*, 2 vols. (Edinburgh, 1842), i, 273–6.

sustained success was achieved in fishing for herring in the lochs of the North West at any time in the first three-quarters of the nineteenth century; there were, it is true, years of legendary success but they were occasional and certainly shortlived in their occurrence in any particular loch.[1] For most years, in fact, the effort and results achieved by the thousands of occasional fishermen up and down the coast were out of sight, not sufficiently notable to be reported and yet still contributing something to the incomes of a great number of people. It is evident that the habit of an occasional herring fishing was not diminishing. Most townships had at least one boat to call into use when the local opportunity came; but, to judge by a survey made in the Loch Carron and Skye fishery district in 1862, not more than one family in three would have direct shares in the available boats.[2] There had not, at that date, been any improvement made in the gear which the typical fisherman would use. Nearly all boats had some nets and the best-equipped had as many as twenty, but many had less than ten and some as few as four. The 2,590 fishermen of the district were from 1850 to 1855 served by a number of curers varying, year by year, between fourteen and thirty-three, while the cure ranged between 1,181 and 4,236 barrels. The curers' businesses within the district were normally very tiny, and out of 114 annual returns made for individual firms on only three occasions did any one business top 500 barrels. Forty-one out of the 114 annual totals produced by individual firms were for less than 100 barrels.[3] This shore-curing resulted in a production per fisherman which ranged between one and four barrels—an income at the most of £2 per head.

II

The rise of the Caithness fishing after 1800 had its repercussions along the northern and western seaboards. Herring fishing at nearby points on the east coast was beginning to offer a stark contrast, in both its intense activity and scale of reward, with the struggling herring fishing of the west; and it seemed a natural ambition for the west coast man to acquire a share in the gains of the booming area. By 1815, partly as a consequence of the recent slight enlargement of the boats, the individual

[1] *Committee on the British Fisheries, 1785*, 13, 82, 96, 104; *Report to the Board of Supervision, by Sir John M'Neill, G.C.B., on the Western Highlands and Islands*, P.P. 1851, XXVI, App. A, 84, 88, 60; *Reports from the Select Committee appointed to inquire into the condition of the Highlands and Islands of Scotland, and into the practicability of affording the proper relief by means of Emigration*, P.P. 1841, VI, QQ. 46, 48.
[2] FBR AF 27/43 Loch Carron, Statistics Record Book.
[3] FBR AF 27/18–36 Loch Carron, Herring Coast Fishing Book.

participant in the summer fishing at Caithness might expect to make at least £10, with the chance of considerably more, while the loch fisherman of the west did very well to make £5. The season, too, was one which fitted very well into the basic rhythm of work on the small farms on which the fishermen of the North West would continue to depend.

However, it was not easy to switch from an occasional fishing in a local loch to a regular participation, for six weeks or two months at a time, in a fishing on the other side of the country. One great difficulty was that of equipment. By 1815 the boats used on the Caithness grounds were at least of 18-foot keel and drifts consisted of at least fourteen nets. The cost of a fully equipped boat was then approaching £100;[1] by the 1840s it was over £150.[2] This contrasted with the £10 which was enough to equip a crew for the loch fishing. The higher standard of plenishing became obligatory for the east coast fishing because the curers (to whom west as well as east coast crews had to turn) were making engagements depend on the boats being up to standard. It was also probably less easy for a crew from the North West, living in an isolated community where there were neither merchants nor curers, to borrow money for gear. The curers who ran the east coast fishing were men of a single interest and not likely to come into contact with or become involved in the complicated transactions of a poor community in which the sale of grain and groceries was interwoven with the supplying of gear and the purchase of agricultural produce; on the other hand the merchants who knew the west coast communities were men prepared for a trade of intricate details in many commodities, and for such men to invest in a boat suitable for an east coast fishing to be made by a west coast crew was to venture money on an unknown prospect with no direct gain in the produce accruing from the investment.

Yet some crews did manage to acquire the boats and the gear which would give them an active place in the fishing of the east. Mainly it was in those parts of the North West nearest to Caithness that the first signs were to be seen of a move to a more concerted effort at fishing. Along the north coast and for a short distance southwards along the coast of Sutherland, a bigger style of boat was adopted by a minority of crews. The parish of Edderachylis, for example, in 1839 had twenty-four large boats, apparently scattered through its coastal townships,[3] They all made the annual trip to Caithness and accounted for a catch of £720

---

[1] BFS, 'Extracts from rental, etc., of Pulteneytown', 1815.
[2] *NSA*, XIII, Banff, 236; Caithness, 101, 153.
[3] *NSA*, XV, Sutherland, 131.

value which would be shared among about one quarter of all families as compared with the £623 taken in the smaller boats, much more widely shared and giving a return of £7 per boat. Even as fishermen, then, the men who had shares in big boats were a modest élite, very much in the minority and scattered around the townships rather than concentrated in any one main centre. In 1840 there were no signs that beyond this north western corner any effort had been made to equip crews for the east coast fishing. For the vast majority of the people, fishing was at best an occasional activity, a sudden and sporadic effort, when the herrings were in the loch and when the small, ill-equipped boat was prepared to catch them.

The rise of the summer fishing on the east coast had another, and at first more important, effect on the livelihood of the people of the west coast. This arose from the employment offered to temporary migrants from the west, during the summer fishing, on east coast boats. At the beginning of the nineteenth century it was not unusual to find two or three hired men in a Caithness crew. When the crews from the Moray Firth and farther south began to fish for herring in great numbers they also hired outsiders to make up the crews. By the 1840s probably a majority of the fleet, then consisting of about 2,500 boats, would have some hired men in their crews; and of these men the majority, although not all, were from the west.[1] Probably between 3,000 and 5,000 men from the west coast were being hired for the season by the boat-owners of the east. A wide area supplied this labour. In some districts, it appears, some townships would send a majority of their men-folk and nearly every family would have some representative away at the east coast fishing; in others the proportions were lower.[2] But most of the areas to the north of Skye, including that island itself, had begun to draw an appreciable part of their income from this annual migration. It was customary for the men to tramp overland, arriving at one or other of the ports of the east in the chance of engagement. Wages were the subject of bargains made just before the beginning of the season. Some individuals had well-established reputations as good workers which allowed them to make better bargains with their employers than did their fellows. But often the arrival of men seeking work, without clear indication as to how many were needed at any particular centre, meant

---

[1] Thomson, *Value of the Scottish Fisheries*, 13–36; *NSA*, XV, Caithness, 101.
[2] *S.C. on Emigration, 1841*, QQ. 3245–7; *M'Neill Report to the Board of Supervision, 1851*, pp. ix, xxviii; App. A, 92; *Report from Her Majesty's Commissioners for inquiry into the Administration and Practical Operation of the Poor Laws in Scotland*, P.P. 1844, XXI–XXIV, App. II, pp. 298, 428.

that some would not find work at all; or men would be scarce and owners would find it difficult to make up their crews, so that wages would rise.[1] There were many uncertainties, then, surrounding the earnings of the hired man. He might find it difficult to obtain any work at all after his long trek, and the wage at which he engaged was subject to fluctuations with part of his income being tied to the catch made by the boat. Year by year, however, a considerable income was transferred from east and west; always some thousands would gain employment and thus generally make individual earnings of between £3. 10s. and £6.[2] On many crofts and for many families it was the biggest source of money income.

Women, too, went from west to east for the summer fishing season, to seek work in the yards as gutters and packers. It was customary, as we have seen, for curers to engage a crew of three women for every boat on his list, and the number of seasonal jobs for women in the yards increased from about 3,000 when the fishing was confined to Caithness to nearly 9,000 along the whole coast in the 1830s. About half of those engaged were local women, the families of the fishermen or of residents in the surrounding districts, and the rest would come from the west along with the men-folk seeking work on the boats.[3] The earnings of the women gutters were much more uncertain than those made by men on the boats: a small sum was paid on engagement but the greater part of earnings consisted of payment for each completed barrel, and this rate (like every other arrangement for the fishing) was the subject of bargaining before the season and could fluctuate somewhat from season to season. But much greater was the fluctuation due to the uncertainty of how much fish would come into a yard for curing. The earnings of the women were strictly determined by the size of the catches of the boats connected with a particular yard—a fluctuation which was probably greater than that seen in the average for the fleet as a whole. In a good season, a woman gutter might take almost as much as would a male hired hand, but in a poor one her earnings might be less than £2.[4]

Early in the century the influences stemming from the great and growing fishing in the east were slowly penetrating along the west coast, drawing eastward more and more people from a population which continued to depend mainly on the working of small holdings of land. But

---

[1] *N. Ensign,* 21 July 1852; *John O'Groat Journal,* 18 July 1851.
[2] *S.C. on Emigration, 1841,* QQ. 3246–7.
[3] Wilson, *Voyage round the Coasts,* ii, 160; *S.C. on Promissory Notes, 1826,* 81.
[4] See above, p. 54.

from 1840 the influence of east coast curers and the east coast boats with their relatively high earnings of the crews suddenly became much stronger in the west. A section of the east coast fishing industry was transferred in full working order to the west when curers began to make arrangements for an early summer fishing to be conducted in the Minch, from May to early July, with bases on the eastern side of Lewis. This was unlike any previous fishing known in the North West because the positions chosen for curing were predetermined, the prices paid to the fishermen were guaranteed and the fishing in the open water was much more certain than that possible in the lochs. The object was the curing of herring mainly for the Continental market and curers, making competitive offers to secure the services of local boats, had to make the same sort of calculation in their offers as they did in their home areas. They had, it is true, to pay a slightly higher freight on herring going to the Baltic than when they exported from the east coast but, because this was an early season fishing, the prices at the German end tended to be higher. Thus the rates they paid to fishermen were only slightly lower than were customary on the east coast and were very much higher, as well as more certain, than anything that had for long prevailed among the lochs of the North West. Engagements were in the standard form, by which herring up to a seasonal maximum for each boat would be accepted at a fixed price. West coast crews with reasonable equipment, if generally less acceptable than the east coast men, were nevertheless given the chance of fishing close to home for a reward likely to be much above that from the loch fishing.

For the curers, once bases were established in the North West, the area seemed to offer a new reservoir from which they could draw on the services of additional crews; for the people themselves, the very proximity of the bases made the opportunities of this novel herring fishing more immediate. Yet there were still difficulties obstructing a mass participation by west coast crews in herring fishings in the open water. With every decade it became more and more expensive to acquire boats and gear up to the minimum accepted standard of the day, although many west coast crews bought at second-hand from the east coast crews out of a stock which was all the greater because of the frequency with which craft there were being replaced. The financial problems of the crofter who was ambitious to obtain a boat had not eased with time and still less was the cottar, the man with no settled holding of his own, likely to be able to meet the expense of equipping a boat for the herring fishing. Indeed, the whole population had the greatest difficulty in rising to the minimum requirements of paying the rent, meeting the food-bill for the deficiency of subsistence supplied by the croft and purchasing groceries and manufactured goods. The kelp industry, which had made a considerable contribution to money incomes, in particular to those of the islanders, collapsed by about 1830. In the meantime the increase of population proceeded and, with every family ever seeking to acquire or keep its personal holding of land, the crofts tended to become more and more fragmented, either by the action of the proprietor or by illicit subdivision. Till mid-century the price of cattle remained in the trough to which it had sunk after 1815 while the continuing spread of large sheep farms, at the very least, restricted the stocks of cattle which could be kept by the smaller tenants. Thus, a typical smallholding of two or three arable acres, with the right of grazing five cattle, provided less than full subsistence and less than enough money to pay the rent. Clearly, there was no surplus here which could be put into an expensive boat. The later forties brought the added difficulty of potato blight and outright famine. All available stocks of money and even of cattle were spent or sold; arrears of rent accumulated; debts were incurred to pay for imported food. For years to come it was going to be difficult to recover the old earning capacity of the croft since cattle stocks once depleted could only be replaced with money which was desperately needed for other purposes; a great mass of debt had to be cleared off before any money could be saved and put into productive fishing ventures. The irony was that, when it was more than ever necessary to earn by the fruits of the sea, it was less and less easy to acquire the means of profitable fishing.

Nevertheless the curers were undoubtedly helping the small-holders to acquire the large boats that created for the crew the possibility of sharing in the east coast and Minch fishings.

At about 17 lads first go to sea, and after three or four years' experience several of them obtain from a curer a boat and nets on becoming jointly bound to fish for him at market prices (or something less) until the boat and nets are paid for. Owners of property they now feel entitled to marry. . . . Their fortunes, however, are bound up in their boat, their shares in her are not saleable and if they have not fallen heirs to their fathers' crofts they probably become squatters . . . before the original debt to the curer has been paid off.[1]

By 1856, not only the west coast of Sutherland but also the northern part of Ross had a quota of boats of the first class (that is, of more than 30-foot keel).[2] In all, in this region there were 125 such boats, within a

[1] *Fourth Report of the Commissioners on the Employment of Children, Young Persons and Women in Agriculture*, P.P. 1870, IX, 332–3.
[2] FB Rep., *1856*, 24–39.

population of about 25,000; that is, in general terms, about one family in seven would have a share in such a boat. For the mainland these figures probably represent the highest level of participation which was to be achieved in the open-sea herring fishing and by 1878 the number of large boats had shrunk considerably with the number of first-class boats standing at ninety-seven.[1] But decline on the mainland was accompanied by rising activity by Lewismen in their local Minch fishing; the number of boats of the first class multiplied in eastern Lewis while it declined on the mainland and by 1878 the district had as many as 182 of them.

This form of herring fishing, tied to fixed stations in Lewis and Barra and capable of being extended by a sojourn at the east coast summer fishing, did not bring the west coast crews to a position where fishing provided a full livelihood. The rewards were bigger and more certain than those of the loch herring fishings, but they remained—on the whole— well below those of east coast crews working in similar circumstances. Partly this was due to the westerners' subjection to the curers. Boats were bought wholly or partly by curers and crews had to work off the consequent debt before they were free to bargain about the price of their fish; then they were tied to the one man and they usually made less than the generally prevailing price for their fish. And the relation of fisherman to curer was not confined to the supply of boats and gear and the purchase of fish; curers also had to become general merchants and to supply their fishermen with a wide range of trade goods:

The fishcurers are generally most liberal in furnishing their fishermen who are not able to procure for themselves with well-equipped boats of the value from £100 to £120 each, also with money and meal in the expectation that in the course of a few years all shall be refunded.

The curers are also ready to make small advances year by year to the crews of boats knowing that while the crews are indebted to them they are bound to fish for them on their own terms, otherwise accounts must be squared.[2]

Thus, in addition to the long-term state of indebtedness on account of boat and gear, there was a seasonal ebb and flow of debts as advances were made at time of need and repaid when the funds became available. In a good season the crews would clear themselves at least of the seasonal account, although a boat would not likely be paid off by the proceeds of a single season.[3] The core of the debt tended to remain even when the fishings were going well and in a bad season even the short-term debts

[1] FB Rep., *1878*, 23.
[2] *Commission on Children in Agriculture, Fourth Report, 1870*, 330.
[3] Ibid., 322–3

would not be cleared. A further consequence of this system was that the fisherman tended to obtain his supplies from the one store and there was at least the possibility of the merchant-owner making heavy concealed interest charge by manipulating prices. In the main, however, the interest of the curers was the purchase of herring at as low a price as possible and it is doubtful whether they were much concerned with the subsidiary gains of the general trading account.[1]

North-west crews not only received less than the full price for the fish that they landed but also tended to end the season having made smaller landings than the east coast men who were fishing at the same stations. For this also the back-log of debt and poor equipment was to blame. West coast men had never had the means of buying, new, the latest and biggest boats so that, compelled to buy at second-hand generally from the east coast men, they found themselves with craft which would generally fish at a lower average than the newer and bigger boats. And a low average catch, coupled with the poorer prices on account of their subjection by debt to the curers, created an income out of which it was impossible to make savings for any improvement. They were caught in a vicious circle of poor yields arising from poor equipment, and of poor equipment enforced by low yields and low incomes; of debts bringing low prices and low prices enforcing further borrowing.

While a minority of north-west crews was breaking into the large-scale herring fishing, the growth of the main east coast fleet at its own summer fishing and the widening of its operations had probably more powerful effects on the crofting population as a whole. The curing stations on the Long Island needed women labourers in the established ratio of three women to one boat. Most of these women were local; they would collect from the townships to work in the yards, being housed by the curers in temporary wooden huts. When ultimately 1,400 boats were engaged in the Long Island fishing some 4,000 women from the island crofts were finding seasonal employ.[2]

The gains during the early summer season—from May to June— would be smaller than would be made in the main summer fishings, for the average take was generally well under 100 crans.[3] The growth of the fleet collected for the great summer fishing on the east coast itself was of an equal consequence to the people of the west. Firstly, more and

[1] *Committee on Crofters and Cottars, 1884*, QQ. 14419–20, 14787, 15174, 15314, 15319; *Commission on Children in Agriculture, Fourth Report, 1870*, 330.
[2] FB Rep., *1887*, 21.
[3] *Sea Fisheries Commission, 1866*, QQ. 32212–4.

more hired men were required, both because the number of crews was greater (reaching 4,000 by 1870) and because in these units a higher proportion of the crew would consist of hired men. Thus, employment for west coast men and women reached its highest level about 1870. Thereafter the number of boats tended to decline somewhat, although in the bigger craft the crew would be increased from six to seven men. The average and the total gain, both of women and of hired hands, increased through the next fifteen years. Average catches increased in the seventies and even the wild fluctuations of the early eighties did not pull down the high average level, taking one year with another. The earnings of the women would increase in almost strict proportion to the increase of catch, and the earnings of the men would move upwards in rather less exact proportion to the catch. Thus by the eighties £30 was being spoken of as the possible reward of six weeks at an east coast port, although the remuneration might well fall below £10.[1]

III

Some of the fishermen of the west had a different object—the catching of the great fish, and in particular cod and ling. This line fishing was pursued in a different way, gave different results and rested on a different social organization than herring fishing. The catches were made on particular banks of prolific yield generally situated in the open water, and to fish for them meant a voyage of ten or twenty miles offshore, whether into the Minch or into the open Atlantic. The white fishers were to be found in relatively few communities sited at points of easy access to known cod banks and had this fishing as their predominant interest; yet not all communities with favoured positions would necessarily engage in such fishings and there was a strong element of tradition in the particular specialization of the cod-fishing community. The main areas of the North West which had developed such traditions were in Barra, in north-west Lewis, in west Sutherland, along the western southern coast of Skye, in Gairloch and among the smaller isles of the southern Hebrides.

The fishing was pursued in open boats, usually of about 20-foot keel and therefore somewhat larger than the all-purpose boats that were used for the occasional herring fishing of the lochs. The crew consisted of about six men, all drawn from the one township and usually sharing equally in the proceeds as owners. Several of the cod-fishing areas suffered from lack of good sheltered water and the boats had to be hauled over open and often stormy beaches at every landing and

[1] *Report on the Lews,* P.P. 1888, LXXX, 3, 30.

launching.[1] Such conditions set a limit to the size of boat which could be used. The method of capture was the great-line. The fishing was a strictly seasonal one but, during the period from January till possibly July, operation was continuous as far as weather allowed. Boats would remain at sea for periods of up to three days. One main weakness of this fishing was its vulnerability to the weather; boats might be kept ashore for long periods during the fishing season and the number of completed trips was fairly small. Nevertheless the results of the season's fishing usually averaged out within reasonably narrow limits. The yield was modest but steady, and the great-line fisherman (to a much greater extent than the herring fisherman) could base his livelihood on the sea. The season's catch could amount to between 1,000 and 2,000 fish of a value between £18 and £36.[2] At least some of the white-fishing communities had considerably less land per family than was usual in the whole region; yet none became completely specialized as fishers and the routine of fishing interlocked with the seasonal requirements of agriculture.[3]

All the fish were cured by drying—a process which we have seen could be performed by the families of the fishermen. Sometimes, then, crews would have all their catch processed by the members of their families; and the finished product might be taken directly to a main market, so that the fishing crews had no local contact with merchant or tacksman. When the men got the resulting cash into their pockets they acquired the means of equipping themselves for further fishing and sustained their independence. Such trading appears to have been the custom of the Barra men, but in other districts the crews had to call on the aid of outsiders and found themselves fishing under much more restricted conditions.[4] Perhaps significantly, in the areas farthest from the market, merchants and tacksmen took a strong grip. In Lewis, for example, merchants and tacksmen provided boats, gear and subsistence for the fishing crews, the merchants controlling the area around Stornoway and the tacksmen the more outlying districts.[5] The grip exerted by the merchant derived from his provision of necessary capital and he used it to ensure for himself the right of purchasing at

[1] *Report on the Lews, 1888*, 3.
[2] R. Fall, *Observations on the Report of the Committee of the House of Commons appointed to inquire into the State of the British Fishery* (Dunbar, 1786), 40; *M'Neill Report to the Board of Supervision, 1851*, App. A., pp. 60, 73, 77.
[3] John Henderson, *General View of the Agriculture of the County of Sutherland* (London, 1812), 19.
[4] *OSA*, XIII, 335.
[5] *Committee on the British Fisheries, 1785*, 155.

prearranged price the whole catch already cured by the fishermen. The control exerted by the tacksman was even closer, for he was able to lay down the conditions of fishing when he sublet the land to occupiers who had no other recourse or power of bargaining. Tacksmen, like merchants, provided capital and subsistence in a form of a truck system in which the value of the fish rendered to him by compulsion was set against the goods and equipment provided for the fishermen. Both merchants and tacksmen, so long as they retained ownership of the boats, would take one-seventh of the catch as their own in addition to the profits derived from handling the remainder. Gross takings would be about £30 in the year, leaving a nominal money return of about £4 to each man; but this was the mere balancing item in an accounting in which, as the suppliers of goods, the merchant and tacksman had the power of arbitrary charging.

Cod and ling fishing continued through the nineteenth century as the surviving steady interest of a minority of the people of the Highlands and Islands. In certain townships situated on well-defined stretches of coast it provided a means by which most families secured a modest but steady income to underpin crofting. Nor was there any tendency for change in the geographical layout of this industry. In the 1880s it was still the general areas already mentioned as notable for white fishing around 1800 that maintained their traditional activity—in particular the northern part of Lewis, with special emphasis on the west coast, Barra, parts of western Skye, Gairloch, western Sutherland and certain of the smaller isles. The signs are, too, that within each area the scale of the effort and the numbers involved remained pretty steady.

Within this settled framework there had been some signs of development in the 1860s. Boats were by then being built slightly larger, with a keel length of up to 25 feet, and they were being taken to fish at greater offings of up to, say, twenty miles.[1] Altogether, there was report of increasing activity but the changes were too gradual to cause any fundamental break in the traditional scheme of operation which focussed the participation, under severe conditions, of small groups in scattered communities.

Indeed, in spite of growth of the sixties, cod and ling fishing is still found in the 1880s remarkably unchanged in its condition as compared with nearly a hundred years before. The increase in size of the boats was still limited by the conditions under which they had to operate: in Lewis, for example, boats still had to be dragged over the beach till afloat and then ballasted with stones which the men carried on their

[1] *Sea Fisheries Commission, 1866*, Q. 32836.

backs, wading to the boat in sea-boots.[1] Particularly on the open west coast this often had to be done in perilous surf, and often no launching would be possible because of bad weather.[2] The operations of fishing were uncomfortable and hard, and the reward was limited both by the relatively small boats and by the frequent interruptions to intensive fishing.

Fishermen, too, remained deeply subject to the curers. Mostly, they handed on their catch uncured to a local middleman who was often the only possible purchaser, although in Barra and in Skye fishermen were sometimes their own curers and might indeed carry the product to market where there was some competition between buyers.[3] But subjection to one merchant was much the more common state, particularly in Lewis.[4] Fishermen depended on curers not only because such men might control the only outlet for their fish but also because they were the only suppliers of credit. Boats were normally supplied in the first instance by merchants who had an interest in the purchase of the catch and, while debt on account of the boat lasted, the sale of the catch to the creditor was enforced. Sometimes the curer remained the formal owner until the crew had paid the full price; and with such ownership went a claim to one-seventh of the catch. Equally pervasive was indebtedness on account of short-term advances made for seasonal supplies, not only for fishing gear but also for meal. Fishermen were often so deeply in debt that they finished the season without any money changing hands, since the total return from the sale of fish was taken up in returning the season's advances if not in paying for the boat. It was, of course, a further consequence that the fisherman took all his supplies from the one man to whom he was also bound to sell his catch and that both buying and selling prices might be turned against him.

Altogether, the line-fishing for cod and ling remained a regular provider of somewhat limited income. But it implicated the fishermen more deeply than herring fishing; it was a necessary and unvarying part of their livelihood for men with tiny holdings of land; the whole life of the township, which served also as fishing base, was keyed to the needs of the fishing. Conditions remained severe and there was little possibility of alleviation or of increase of income so long as the basic traditions were accepted.

---

[1] *Fish Trades Gazette*, 19 Apr. 1890.
[2] *Commission on Crofters and Cottars, 1884*, QQ. 16110, 17076, 17241.
[3] *Commission on Crofters and Cottars, 1884*, Q. 26493; FB Reps., *1894*, 183; *1898*, 221; *1899*, 231.
[4] *Commission on Crofters and Cottars, 1884*, QQ. 14671–6, 15174–8, 15314–22, 15534–46, 15899–926.

IV

To the east of the Kintyre peninsula, in the several lochs running
north from the Clyde estuary, from the late eighteenth century onwards
fishing was generally conducted with greater success than elsewhere on
the west coast. Two special ingredients made for this success. Firstly,

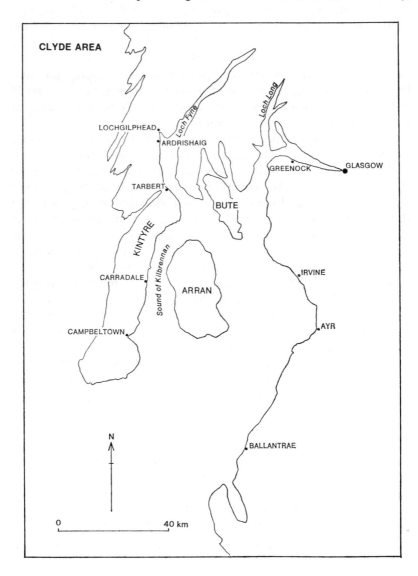

the herring appeared more regularly in Loch Fyne than in any other enclosed water around Great Britain. There were fluctuations, of course, and in any period of ten years the largest fishing could be as much as three times as productive as the smallest; but there was never anything like the complete failure that periodically afflicted every loch in the North West. Secondly, there was a ready and large market near by which even in the days of sailing ships could be reached within a few hours and which would daily take large supplies of fresh fish. Prices therefore tended to be somewhat higher than farther to the north although in the fluctuations of daily supply, with its sudden excesses, not all the herring were at any time sold in the Glasgow market. Helped by favourable conditions of market and supply, the people of Loch Fyne and of the eastern side of Kintyre contrived to draw a large and fairly regular income from herring fishing.

In the late eighteenth century Loch Fyne and its drift-net fishing had 500 to 600 boats drawn from the various communities around the loch; they would collect for a season from July to December.[1] Still farther east, the men of Loch Long and Loch Goil might find fair fishing in their own lochs and then they had even better marketing opportunity than the Loch Fyne men. But their fishing was less certain than that in Loch Fyne and they might move to that less fallible centre for the best of the season. The fishermen of this region were also farmers but certainly they drew more of their income from the fishing side. Their boats were small but, in the conditions of the enclosed loch, this meant no serious deficiency. They were fairly well-equipped with nets which would be set so as to run across the loch, while boats might be so closely packed as almost to jostle each other. The upper section of Loch Fyne seems to have provided the most lucrative fishing area.[2] From here part of the herring would be carried overland to the head of Loch Goil, whence the fish would be taken by ship to Glasgow for daily sale. The boats were owned partly by the fishermen themselves, partly by shore capitalists, although it is not clear whether the latter were curers or merchants. One arrangement was for a landsman to share with four others who made up the crew and take a corresponding fifth share in the proceeds. The return on the season's fishing was about £10 per man.[3]

Drift-net fishing, soundly established before 1800, grew and developed through the first three-quarters of the nineteenth century. Fishing at first was generally combined with the holding of land and industrial earnings, added to the return of the land, gave incomes out of which investment in improved gear was possible without recourse to external

[1] *OSA*, V, 292.   [2] *OSA*, III, 172–3.   [3] *OSA*, III, 434.

sources of credit. The rising level of output and earning opened the way to full specialization in fishing, on the basis of regular and reasonable incomes, giving yet more scope and incentive to the improvement of equipment.[1] In the 1820s half-decked boats were introduced and by the 1840s the bulk of the Clyde fishermen were using such craft, the largest at this time being of 22-foot keel.[2] Such a boat, combined with the 600 yards of net which represented the best gear of the time, would cost £100 and individual incomes of £20 or £30 could be made thereby.[3] The drift-net fishermen were migratory, moving between the different parts of Loch Fyne and the waters around Arran and Bute. A crew might start, for example, in June in Loch Long, then move to upper Loch Fyne, and finally go to the more open water of the Sound of Kilbrennan. If the catch was large the greater part would be cured, but there was also a considerable daily sale of fresh fish.

Such was to be the basic programme of activity of the bulk of the fishermen of the region for the next three decades. Until 1870, each successive decade saw on the average more boats gathered annually in Loch Fyne—from 300 in the period 1827–36 to 558 in the period 1857–66.[4] The marked tendency was to build the boats ever larger and to equip them with larger drifts of nets.

In the 1830s the drift-net fishermen found, or felt, themselves threatened by a rival body—the fishermen who adopted the technique variously described as 'trawl-', 'seine-' or 'ring-netting'.[5] As a description, 'ring-netting' is probably the most accurate term. In this new form of fishing—new, that is, to the area—two boats operated together, holding the different ends of a net some 150 yards in length; the net was cast as one boat moved from a starting-point near its partner, to form a wide circle as it returned to the original point, and the haul was completed by gradually constricting the circle till the fish were in a concentrated mass ready to be hauled aboard.[6] This method was well suited to the sheltered waters of the Clyde estuary and was soon seen to be providing hauls considerably greater per man than those which fell to the drift-net fisherman. It had the added advantage that the equipment was considerably less costly.[7] The result was not that the

---

[1] *NSA*, VII, Argyll, 450–1. [2] *NSA*, VII, Argyll, 32.
[3] *NSA*, VII, Argyll, 368. [4] *Report on the Herring Fisheries, 1878*, 424.
[5] *Sea Fisheries Commission, 1866*, Q. 53345; *Report on the Herring Fisheries, 1878*, 405.
[6] *Sea Fisheries Commission, 1866*, Q. 53708; *Report on the Herring Fisheries, 1878*, 419.
[7] *Sea Fisheries Commission, 1866*, QQ. 50352–3, 51803, 51978–9; *NSA*, VII, Argyll, 264, 368.

new form of fishing replaced the old but that embattled groups argued for and against the prohibition of ring-netting. The annual income from the newer method was definitely higher at £80 to £120 per man than from drift-netting at £40 to £50.[1] The competition between the techniques did not extend to a direct struggle for the fishing grounds for they tended to take up different stations: ring-netting was carried out in bays too shallow for the drift-net. While the main congregation of drift-netters was in the upper part of Loch Fyne, ring-netting was confined to the lochs and open waters below Otter Point. There were, too, some seasonal differences, with ring-netting continuing through the winter. It was possible, indeed, for ring-netters also to carry drift-nets which they would shoot on occasion.

The true points of conflict concerned the effect on the market of the supply from the ring-netters. Prices were on the whole lower because of this extra supply and they were made more uncertain by the fluctuations specially inherent in the ring-net method. But it was presumably the fear about the effect of ring-netting upon fish stocks that most moved outside opinion. The climax of the argument came when, in 1851, trawling was prohibited. The regulation seems not to have been effective for it was strengthened in 1861. However, the Sea Fish Commission of 1864, heavily committed to freedom from all regulation, recommended that the prohibition should be rescinded and in 1867 the loch was opened once more to trawling. The ring-net men, many of whom had gone over to drift-netting in the short period of effective regulation since 1861, speedily returned to their original specialization and by the seventies there were once again more than 100 ring boats at work in the loch. The picture of herring fishing in Loch Fyne and the lochs and sounds to the south and east was one of a slowly growing annual complement of boats, always composed mostly of drift-net boats. But, apart from the years of prohibition between 1851 and 1867, the ring-netters always accounted for a considerable proportion of the catch because of their superior capacity. On the whole, even when ring-netting was banned, the catch was on the increase. There were severe fluctuations, but always above a minimum level which was raised in each decade until the mid-sixties.

The growth of the catch, both from drift-nets and from ring-nets, no more than matched the increase of the market. The industrial population within immediate reach was growing rapidly in the middle of the nineteenth century, and more and more could be sold in the daily market for fresh fish. The connection between boat and market was much

[1] *Sea Fisheries Commission, 1866*, QQ. 63671, 63679; *NSA*, VII, Argyll, 368.

improved when steamers began to ply, daily, between the fishing areas and Greenock and Glasgow. Six or seven vessels would be standing by as the night's fishing was brought in.[1] With fish from ring-net boats the transports could be in Glasgow by early in the day although the supply from the drift-net boats, slower to unload, would not arrive till later in the day. Not all of the fish sold in the Glasgow market was consumed there. Steamers linked with the railways and enabled fish to be sold in Billingsgate within fifteen hours of its landing in Loch Fyne. Fish were generally sold by the boats' crews to the merchants running the steam vessels on the basis of competitive offer and the price might fluctuate considerably through the day.

The growth in the market for fresh fish did not entirely obliterate curing at the site of the fishing. In the eighteenth century about two-thirds of the catch had been cured and, while this proportion probably diminished, the absolute amount cured from year to year did not decline. The custom was to sell fish for the fresh market as long as price held above a certain level; below that point the curers would buy for pickling. In practice this meant that the first fish of the day would all go away fresh.[2] Sometimes the daily landings would be so small as to keep the price above the critical level; but if the day's supply ultimately proved too great for the immediate market then the curers stepped in to take the excess. The system, in other words, ensured a fairly regular supply to the fresh market, with gluts averted by the operations of the curers; a price above the curers' level would prevail for much of the catch; but, when the supply became too great, buying by the curers acted as a platform which prevented the absolute collapse of price. Curing was carried out mainly on board vessels fitted out for the purpose.

Herring fishing within the Clyde estuary always gave a good and reasonably steady return for a smallish capital outlay. Drift-net boats and their gear were increased in size and the drift of nets was somewhat lengthened so that by 1864 the minimum cost of such equipment was £200.[3] But this was somewhat low, relative to the amounts that an east coast crew had to expend to be able to fish effectively in the open sea. The capital costs of ring-net fishing were lower still, for a net shared by two boats cost £20 to £25 and the boats themselves ran to only £24.[4] From such boats, whether used for ring-net or for drift-net fishing, individual incomes of between £40 and £100 per year were regularly earned, with the ring-net usually providing the greater weekly

[1] *Report on the Herring Fisheries, 1878*, 422–3.
[2] *Sea Fisheries Commission, 1866*, Q. 5308.
[3] Ibid., Q. 34060.   [4] Ibid., QQ. 33994–8.

rate. Thus, the fishermen were able to retain their independence. Boats were owned, although often in unequal fractions, by members of the crew; two or three men might own the whole boat and its gear but the necessary money all came from within the crew. Nor do they appear to have run up debts to hold this position.[1]

With rising, although still moderate, capital costs and with the possibility of earning adequate incomes, the tendency was for fishermen to operate full-time. In the upper parts of Loch Fyne the fisherman-crofter remained typical but farther south, particularly along the eastern side of the Kintyre peninsula, land ceased to be an interest. One difficulty in full specialization was that the herring fishing did not occupy the whole year, even for the crew prepared to move from one part of the region to the other. The winter and early spring months might be spent in great-line fishing but, when a formal close season herring fishing was declared from January to May, bait for the lines—previously obtained by a small-scale herring fishing—became scarce. Some men, then, were driven back to the cultivation of potatoes.

[1] Ibid., QQ. 52765, 53357, 57450–4.

Crofting and Fishing: the Northern Isles 1800–1880

I

BOTH THE general and the nineteenth-century fishing experience were remarkably different in the two groups of northern islands, Orkney and Shetland. Indeed, Orkney, surrounded by waters much fished by incoming fleets, was strangely lacking in any determined fishing tradition. The feebleness of the fishing effort was particularly noticeable towards the end of the eighteenth century. Boats there were in plenty, owned by the small farmers who lived by the long coastlines of the islands, but they were small and were used in bursts of casual endeavour, almost entirely for the catching inshore of a subsistence food supply: 'Every individual during the summer and harvest months, has a seat and share in a fishing boat and catches fish for his house and family.'[1] At no season, then, did the Orcadian farmer with his boat remove himself from farming for a continuous period of fishing. Boats in Rousay, for instance, were so small as to be valued at only £3 each.[2] Virtually the only commercial fishings were for cod by two small groups in Stromness and in the parish of Walls, for dog-fish in Orphir, and over a wider area for lobsters. The cod-fishing from Stromness for which six boats were fitted out, was of great-line type.[3] A crew, owning its boat, would divide annually about £21, which meant less than £4 each; when they did not own the boat, one-third of the gross went to the employer and the crew received less than £15. Twelve Walls boats followed a similar fishing.[4] As for lobster fishing, there were said to be sixty boats engaged in Orkney[5]—tiny craft of no more than 12-foot keel and each of the value of £6, giving an annual return of £7 per man.[6] Twenty-four boats in the parish of Orphir were fitted out for the catching of dog-fish, by handline, at a short distance offshore:[7] the value of this fish lay partly in the oil which was extracted from the livers, but they were also cured and smoked and then sold for 5s. per dozen. Thus the total of boats engaged in any form

---

[1] *OSA*, V, 41. [2] *OSA*, XII, 338. [3] *OSA*, XXI, 435. [4] *OSA*, XVII, 319.
[5] Ibid. [6] *OSA*, XIX, 399. [7] Ibid.

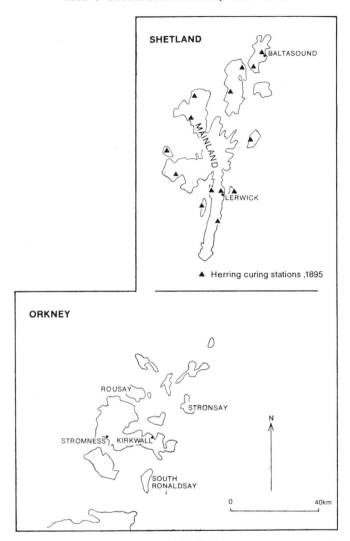

Source: *Shetland Times.*

of commercial fishing in Orkney was tiny and the individual earnings added only small amounts to the farming which was the main occupation of almost all inhabitants. The main supplement to farming income, in fact, came from kelping and the development of this industry in the later eighteenth century was generally held responsible for the neglect of fishing.

Fifty years later the picture was different. Kelp had declined and fishing had grown as the part-time activity and source of monetary gain of the farmers. The development was greatest in herring fishing and in Orkney as a whole there were then said to be 724 boats used mostly for this purpose.[1] The herring boats, which were widely distributed through the islands, were much larger than those of the 1790s; in North Ronaldsay, for example, they were of 28-foot keel[2] and were thus larger than any boats being used by the farmer-fishermen of the west coast. The crews, all of them part-time and drawn from all over the islands, moved to the central points of operation (particularly in Stronsay) and fished consistently for the summer season from late July to September.[3] The fish were landed for shore-curing, being purchased by the curers at about 10s. per cran; an Orkney fishing station had the appearance and vital action of a curing station on the mainland.

Thus, about the middle of the nineteenth century, Orkney seemed to be moving along the road of improvement by which the men of the east coast had so increased their catching power. Boats and equipment were increased and drifts of nets were, by 1864, up to twice as extensive as they had been half a century earlier. But the effort had an insubstantial basis: the fishermen, in part farmers and engaged only in the local fishing, failed to produce enough even to justify the initial investments, let alone allow further improvement or full specialization by their owners. The annual average catch was always less than 100 crans and often less than fifty, which scarcely allowed the proper maintenance and replacement of boat and nets originally costing £150.[4] Thus the number of boats began to fall; in all, 360 were fitted out for the summer fishing in 1855 and only 220 in 1865.[5] In the meantime, debt to the curers became a serious problem.[6] With time, a smaller and smaller proportion of the crofting population took any part in the fishing. In part the decline was due to the absorption of effort in agricultural improvement; after all, a growing proportion of the population were full-time farmers, and labourers and much of the available capital went into the land.

By 1914, then, there was left in Orkney only a small band of fishermen who were commercially engaged and a larger number who made a desultory effort at finding some subsistence with their small boats. The true herring fishermen were now only to be found on the island of

[1] *NSA*, XV, Orkney, 214.   [2] Ibid., 111.   [3] Ibid., 163.
[4] *Sea Fisheries Commission, 1866*, QQ. 31441, 31519, 31523.
[5] Ibid., Q. 31548.
[6] Ibid., *Commission on Crofters and Cottars, 1884*, Q. 24883.
[7] *Sea Fisheries Commission, 1866*, QQ. 31322, 31325, 31389, 31454.

Burray, with a tiny number scattered around some of the other islands.[1] For the rest, line fishing was carried on by some particular groups. There were, for instance, fifty fishermen in Kirkwall, apparently fully dependent on fishing, owning twelve boats of 35-foot to 45-foot keel and catching cod to sell either to curers in the outlying stations or as fresh fish in the market at Kirkwall. There were line fishermen, too, in South Ronaldsay who sold small cod to the Kirkwall curers. Rather different were the Stromness fishermen who used a small yawl-type of boat to catch haddocks in spring and lobsters in summer. Their incomes were reported to be as high as £3 to £4 a week during the season but there was no winter fishing for them with their small boats. In aggregate, fishermen by 1914 comprised only a tiny fraction of the total population of the islands.

II

Shetlanders had a very different tradition and were able to adapt their fishing effort—which was always of considerable scale—to the changing opportunities of the eighteenth century. This greater vigour of response goes back to traditions of fishing which were established as early as the seventeenth century. By 1700, a large proportion of the Shetland population was involved in an intensive seasonal fishing; as the century progressed they tended to venture farther and farther to sea and to draw more and more of their money incomes from the sea.

Much of the fishing was intermittent and local, being conducted in the voes or in tidal races between the islands with rod or handline. This was essentially a minor by-employment of men when they were staying at home in winter or in the heavy agricultural seasons of spring and autumn. Some of the fish caught, such as young saithe, codling and haddock, were used for immediate subsistence, but the larger saithe and cod had a commercial value since the oil of the liver was extracted and sold; so plentiful was the local supply of fish that the flesh of the fish would often be thrown on the dunghill. Local fishing of this sort was the almost universal occupation of the Shetland farmers. All through the year, when they were at home, the tiny boats would be launched, particularly in the evenings.

The main commercial fishing, however, was 'the haaf', that is the deep-sea fishing for cod and ling, to be cured and exported from the islands. These great fish were sought directly to the east and to the west of the islands, on the grounds which—at least to the west—extended to

---

[1] *Report of the Scottish Departmental Committee on the North Sea Fishing Industry,* 1914, *XXXI*, App. 10, p. 213.

the edge of the continental shelf. The earliest efforts at exploiting these stocks had been confined within short distances of the shore; but during the eighteenth century the boats were taken farther and farther to sea, especially on the western side.[1] By the end of the century grounds more than forty miles out were exploited by the small open boats. It was important to work from stations as near as possible to the distant grounds and therefore sometimes at some distance from their farms. But there was another reason for establishing fishing bases that were distinct from the agricultural settlements where the Shetlanders cultivated their smallholdings through most of the year—the curing of the cod and ling required the use of a wide stony beach. Thus the haaf stations, occupied only during the fishing season from May to August, were strung around the extremities of the islands, being particularly numerous around the north-western rim, along the middle west coast, on the south-eastern tip and in the relatively far-flung islands off the east coast of the mainland.[2]

The annual move to the haaf stations was an organized and large-scale operation; conducted under the control of the landlords or of the tacksmen who secured the necessary powers over the fishermen-tenants by a leasing arrangement with the landlords, it was designed to take the crews away from their agricultural bases so that they would devote themselves to fishing and nothing else in situations where an adequate facility for curing could be provided. Each estate would have a haaf station to which its crews—made up from its tenantry—would resort, although the laird did not necessarily own the land backing on to the station.[3] Nor was the station necessarily the point of suitable characteristics lying nearest to the agricultural base. Near the chosen beach, in the area just above high-water—a belt 100 yards wide was allowed for this purpose—there was built a cluster of huts which were mere walls of dry-stone, temporarily thatched with turf or heather. Each crew would occupy a hut collectively while the curing staff would also have their accommodation;[4] a larger 'booth' was used for the practical business of curing. At a large haaf station about seventy boats might gather with their crews, or an assembly of about 400 fishermen, to whom

[1] C. A. Goodlad, *Shetland Fishing Saga* (Lerwick, 1971), 101.
[2] Hance D. Smith, 'The Historical Geography of Trade in the Shetland Islands, 1550–1914', Unpublished Ph.D. Thesis, University of Aberdeen, 1972, 108.
[3] Ibid., 108.
[4] Samuel Hibbert, *A Description of the Shetland Islands, comprising an account of their geology, scenery, antiquities and superstitions* (Edinburgh, 1822), 507–8; Seaforth Muniments, SRO GD46 Sec. 13 152/2, 'Report on Shetland fisheries', 1837.

would be assigned a yard staff of about seventy.[1] Some stations, however, were much smaller.

The craft in which the crews had to tackle these ventures into the open sea were open boats of very distinctive design and of two sizes. For the longer journey to the west the 'sixareen' was used, about 18-foot in keel and crewed by six men each of whom would work an oar; the smaller 'fourareen', used on the eastern side of the islands, had four men to the crew and was of 16-foot keel.[2] Both were of a substantially similar design which apparently owed much to Norse influence. The boats had to ride both the short and broken water near to the island coasts and the long rolling swell of the open sea, and they were designed to sit light on the water, drawing no more than two or two-and-a-half feet, with a high flared bow to keep the shallow well dry.[3]

The most arduous and spectacular journey was that to the western haaf. From the stations on the west coast the boats would set out in a concerted move about 10 a.m. of a Monday. Progress to the fishing ground might be aided by a small sail but the great art in managing the sixareen was through the oars; the craft was guided to run the swell and avoid the disaster of being struck broadside. By nightfall, the boats would have reached the grounds, which were well-defined in relation to each station. On arrival, the lines—running to 4,000 fathoms in length and carrying up to 1,200 hooks—would be baited with haddock, herring or saithe, which had been caught by handline, and would be shot in a crew operation in which each man played a different part.[4] After lying two or three hours they would be hailed, an operation more lengthy and complicated than the original shooting. The line had to be coiled as it was returned, and the fish were removed, gutted and decapitated as the boat was kept moving at a rate to allow the convenient, simultaneous performance of the other tasks. In the course of one trip to the grounds the lines would be shot and hailed two or three times, each completed cycle of operations occupying six to eight hours. As a result, on a successful venture, there might be hauled aboard about 140 ling, along with lesser numbers of fish of other types. Finally came the return to the

<hr/>

[1] Hibbert, *Description of the Shetland Islands,* 527.
[2] Seaforth Muniments, SRO GD46, Sec. 13, 152/1, 'Report on the Shetland fisheries', 1837.
[3] Goodlad, *Fishing Saga,* 103–7.
[4] Arthur Edmondston, *A View of the ancient and present state of the Zetland Islands,* 2 vols. (Edinburgh, 1809), 244; John R. Tudor, *The Orkneys and Shetlands: their past and present state* (London, 1883), 135–6; Hibbert, *Description of the Shetland Islands,* 510; Seaforth Muniments, SRO GD46, Sec. 13, 152/3, 'Report on the Shetland Fisheries, 1837; *Committee on the British Fisheries, 1785,* 26.

station, again running the swell, and navigating—for the journey would start out of sight of land—by the feel of the direction of the sea running under the keel. The most dangerous moment was the approach to land. The boats had to find their way through the indistinct clefts in a coast of stark cliff and rock. In mist or with an onshore wind it demanded great skill, and there was always the threat of widespread disaster. At the very best a boat might make two trips to the haaf grounds in a week.

The ling was the main object of fishing and, when the catch was heavy, all other fish brought up in the line would be jettisoned; but if there was room the other fish would be brought back too. Cod would be sold on account to the curer, like the ling; but any other fish were kept by the fishermen, who were also allowed the benefit of all the heads and livers.[2] On return, the ling would be handed over to the curer or his agent; they would be weighed and placed to the crew's account for the final settlement at the end of the season. Then the curing proceeded, by the efforts of a staff that was attached to the station as a whole rather than to particular boats.[1] Fish were cut open from head to tail by a 'splitter' and half the backbone was removed. Then a 'washer' cleansed them of blood by washing in sea-water. After draining, the fish were then placed in a large vat in alternate layers, salt and heavy stones being used to keep the fish in the pickle. After some days they were taken out, washed and piled in small heaps, or 'clamps', to drain again. Then, in the final stage, they were placed on a shingly beach, exposed to the sun. After being alternately clamped and exposed singly, with skin or flesh uppermost according to the particular curing formula, they were built into large stacks or 'steeples' which, in order to equalize the pressure between the particular fish, were taken down and rebuilt with the fish previously at the bottom later being laid uppermost. The building of a steeple was an intricate task. The completion of the cure was indicated by a white efflorescence, and the fish were then carried to a dry cellar lined with wood, if not immediately shipped.

The succession of voyages would continue throughout the three months during which the crews were at the stations, living in the rough lodges when the boats were beached at weekends or when fishing was interrupted by bad weather. Necessarily, with open boats, the weather would often keep the crews ashore and in a season about eighteen trips on average would be made. Finally, at the end of this gap between the 'voar', or sowing on the croft, and the summer-time tasks of peat-

[1] *Committee on the British Fisheries, 1785*, 64.
[2] Wilson, *Voyage round the Coasts, II*, 331; Hibbert, *Description of the Shetland Islands*, 519–20; Tudor, *Orkneys and Shetlands*, 137–40.

gathering, the crews would return to their townships to take up once again the routines of agriculture.[1]

In all this the fisherman was closely controlled by his landlord. Shetland landlords generally had great social power and their functions went far beyond the simple management of land. Concentration in the ownership of land by the late eighteenth century put the greater part of the farming population of Shetland in the situation of tenants of a relatively small group of aristocratic landowners. The estates were frequently fragmented, with different sections—interspersed with land belonging to other lairds—in different parts of the islands; indeed, a single township might well be split between several owners, although each of these might have acres and tenants spreading through many other townships.[2] The landlords alone were able to accumulate capital and to perform the functions that required capital. Thus when the German traders—on whom the population had depended for trade goods as well as for selling their produce, particularly fish—left at the end of the seventeenth century, it was the lairds who had to step in to supply the needs of a population which inevitably had a considerable dependence on external trade.[3] They would commission and even acquire ownership of the vessels that plied to the mainland of Scotland and to the Continent; they became directly involved, through agents, in transactions in Continental markets, they purchased the general wares and the fishing materials needed by their tenants, and they sold the produce of the estate. But, most profitably of all, they began to deal in fish, the commodity on which the payment of their rents very largely depended. Not only did they buy and sell the produce of their tenants but also they began to undertake the curing of the most profitable and prolific type of fish, the ling. Dried ling became the main saleable product of the islands, much of it going direct to the Continent and, in particular, to the North German ports and to Spain.

When the lairds began to lose out in the import and export trade to merchants who did not have the advantage of holding land, they kept their grip on the internal trade and, particularly, on the most profitable part of it, the disposal of fish caught by the tenantry (which involved them directly in the operations of curing).[4] To make their control over curing the more complete, they devised the special fishing tenures by

---

[1] Edmondston, i, *View of the Zetland Islands*, 242; Hibbert, *Description of the Shetland Islands*, 589; Tudor, *Orkneys and Shetlands*, 134.
[2] Smith, 'Trade in the Shetland Islands', 58–9, 107–8.
[3] Ibid., 119; *Committee on the British Fisheries, 1785*, 63.
[4] Edmondston, *View of the Zetland Islands*, i, 295–305; Shirreff, *General View of the Shetland Islands*, 15–16; *Committee on the British Fisheries, 1785*, 63.

which the holding of land was made conditional upon the tenant engaging in the haaf fishing at a station organized by the landlord; and, further, the tenant had to sell all the ling which he caught to the landlord at a price which would be determined only at the end of the season, by dictation rather than by any process of bargaining. Since the access to the fishing stations was controlled by landlords who would let their land only under fishing tenure, all who wished to fish at the haaf had to operate under the strict control of a landlord who then also intervened in the other trading transactions of the tenant's life. The only gaps in this system, which turned the mass of the population into compulsory suppliers of fish to the curer-landlords lay in the one or two estates such as that of Bruce of Sumburgh, who allowed his tenants to sell freely to merchants; and in these (probably few) estates with no involvement in fishing and where the tenant would not have the chance of participating in the haaf.[1] A large proportion of the 3,000 or so men who manned the haaf boats were operating under fishing tenures and only a small number were allowed to sell in the open market.[2] But the numbers who were free of fishing obligation and did not participate in the haaf seem to have been fairly large since crews which aggregated 3,000 members can only have drawn in about half the families of the islands.[3]

To conduct the fishing, individual tenants were linked in the long-term groupings by crew. The crew might be in part a kinship group in which close or distant relatives were associated; but it might also comprise individuals whose sole link was that of being tenants of the same landlord, although usually they would also be drawn from one township. Generally, the members of a crew would acquire a boat as joint and equal owners, sharing the earnings equally. Sometimes, however, they hired boats which were made available, without transfer of ownership, by the landowner.[4] The lines and hooks, too, would be provided in equal quotas by the members of the crew. In fact this ownership was often more nominal than effective. Landlords would provide boats under credit terms, with the provision that as long as the debt was outstanding the creditor would have a direct share in the proceeds; one

[1] Committee on the British Fisheries, 1785, 24, 27, 65; NSA, IX, Shetland, 94; Smith, 'Trade in the Shetland Islands', 216.
[2] Edmondston, View of the Zetland Islands, i, 295–305; J. Kemp, Observations on the Islands of Shetland and their Inhabitants (Edinburgh, 1801), 8; Hibbert, Description of the Shetland Islands, 507, 595–7; Committee on the British Fisheries, 1785, 65.
[3] Shirreff, General View of the Shetland Islands, 15–6, 77.
[4] Edmondston, View of the Zetland Island, i, 235.

seventh of the gross receipts went to the boat and as long as the land-lord had not been repaid he would take this share.[1] The boats were imported from Norway, consisting of pre-fabricated parts which were then locally assembled. In 1800, a sixareen cost (as imported) less than £12.[2] Similarly, lines and hooks would be provided out of the landlord's store, again bringing with it an unresolved debt. Altogether it cost about £25 to provide and fit out a sixareen.[3]

The landlord would generally extend credit to his tenants for other purposes. Tenants, in large measure, succeeded in living self-sufficiently with food from the croft, peat from the hill, the use of local building materials and ingenuity in home manufactures. But there was a residue of goods that had to be purchased. Meal, tobacco, sugar, tea, cotton thread and cloth were occasionally or regularly obtained from the land-lord's or tacksman's store.[4] Thus every tenant built up a complicated account with his landlord. On the one side were the rent, the sums due on goods of current consumption, payments on account of boat and gear; on the other was the value of the fish credited from the haaf and, possibly, of other products of the croft which the landlord would acquire in order to re-sell. The annual settlement would be made in September when the landlord knew the price which he would get for his cured fish and when he would be able to make the 'correct' allow-ance according to 'the price of the country'. The outcome of this settle-ment varied from year to year. Sometimes the tenants would be clear but sometimes debts would be carried over to the following year. It was normal for a tenant's debt to increase seasonally because of his need to keep going through the season until he could have credited to him the proceeds of his work; in a good year this seasonal debt, at least, would be cleared at the end, but to this had to be added a larger debit on account of capital goods which could normally be cleared only after some years. Thus it was fairly normal for the tenant to be in debt to the landlord on this long-term account while the short-term debts, as they stood at the end of the season, fluctuated from year to year.[5]

The size of the Shetland boats and the standard of the gear was not far below that which prevailed in other parts of Scotland at this time, and the fishing on the edge of the continental shelf probably gave as

---

[1] Thomas Gifford, *General Remarks on the British Fisheries* (London, 1784), 49.

[2] *Committee on the British Fisheries, 1785, 26,* 64; Edmondston, *View of the Zetland Islands,* i, 246.

[3] Edmondston, *View of the Zetland Islands,* i, 246.

[4] Kemp, *Observations on the Islands,* 9; Edmondston, *View of the Zetland Islands,* i, 247–8; *Committee on the British Fisheries, 1785,* 26, 65.

[5] *Committee on the British Fisheries, 1785,* 26, 64.

good yields as were to be made anywhere, so that earnings were meagre mainly because of the low level of prices as paid by the landlord. Whether these low prices were the result of the landlord exploiting his position of power was the subject of much discussion. Most of the ling —and therefore of the haaf fisherman's catch—was purchased at the fairly low 'price of the country' which was agreed by landlords acting in concert and which, although it might fluctuate from season to season, was fixed with an eye on the price of cured fish; it was set, in fact, so as to leave an assured margin that the landlords regarded as fair, for every cwt. of fish cured.[1] Thus, a margin of 5s. on every cwt. of dried fish (which sold for less than 20s.) was normal, while the return to the fisherman on the $2\frac{1}{4}$ cwt. of wet fish needed for 1 cwt. of dried fish was less than 10s. It was by this margin on cured fish that the landlord made his greatest gain, being content with a moderate rent if he could be assured of the profits of curing (which indeed were higher than the rent of the land).[2] So much was a Shetland estate valued for the control it gave over a fishing tenantry, rather than for its declared rent, that estates might sell for more than fifty years' purchase.[3] An alternative for the landlord, of course, was to allow a higher price for the fish and to take more in rent out of the tenant's expanded income. A few landlords seem to have operated on this principle of high rents and market prices but it had its most important application with those landlords who allowed the tenants to sell freely to merchants giving higher prices than the 'price of the country' allowed by the majority of the curers. Tenants did not greet the opportunity to sell their fish in an open market as a great advance, partly because of this manipulation of rent, and partly because without the support of landlords in meeting their seasonal needs they tended to fall into debt to merchants and were then left with as little power of bargaining as they had under the thrall of the fishing tenure system.[4]

There is no doubt, then, that the lairds were big gainers by the haaf fishing and that their first interest was to have a large tenant manpower at their disposal. Interest in fishing impinged severely upon agricultural policy; the creation of a multitude of smallholdings to accommodate a

---

[1] Edmondston, *View of the Zetland Islands*, i, 295–305; Patrick Neill, *A Tour through some of the Islands of Orkney and Shetland* (Edinburgh, 1806), 108–9; Seaforth Muniments, SRO GD46, Sec. 13, 152/45, 'Report on the Shetland Fisheries', 1837.

[2] *Committee on the British Fisheries, 1785*, 29.

[3] A North Briton, *General Remarks on the British Fisheries* (London, 1784), 20; Shirreff, *General View of the Shetland Islands*, 16.

[4] See below, pp. 142–5.

growing population was the prime object, and any increase in the size of holdings which might allow tenants to become independent of fishing was obstructed. But it is not certain that the increase of population was 'caused' by the policy of accommodating the people with land. The increase of population which was manifest in the latter decades of the eighteenth century and which continued until 1861 was certainly in line with the interests of the lairds; they did nothing to hinder it or to encourage emigration in face of the increase. But the rate was not particularly notable for the time, and it was less than in those western islands where kelp played a major part. What can be said is that by their land policies, with land made widely available in smallish holdings, the Shetland lairds shaped an environment in which the inevitable resort of the tenant was to combine fishing with farming. But the subsistence derived from the land seems to have been much nearer to the full requirement than, for example, was possible in the western Highlands.

The fishing operations of the Shetland lairds were defended on the ground that they were supplying services—as merchants, as suppliers of capital, as curers and as middlemen in fish—that otherwise would have been at best erratic and uncertain; thus they gave the tenants at least a secure framework in which they could pursue their fishing. After all, the tenant-fisherman had a smallholding of land from which he could extract something close to a full supply of food; he was provided without any question with the necessary implements for fishing; he had the assurance of being able to sell the fish at a price which was roughly known in advance; and in difficult seasons he knew that he would be supplied on credit with the necessities of life. His income was modest but it was, within limits, guaranteed. The question was whether an open market in which the fisherman could have sold his fish to the highest bidder would not have given him a better income; and whether the resources of a market rather than a paternalistic system might not have given all that was needed in the way of credit at short- and long-term. There is some evidence, indeed, that by the end of the eighteenth century the paternalistic system was beginning to crumble somewhat, having to be enforced by hard prohibitions. Some fish was being sold outside estate control to merchants and it obtained a higher price than the landlord's 'price of the country'. Yet it should be remembered that such transactions were usually made at the most favoured sites and that merchants might not have been able to offer such high prices in the more remote stations where most of the fishing took place. In addition, the experience of the nineteenth century eventually showed that when landlords ceased to deal directly in fish and other commodities they were replaced by merchants who might operate as direct and comprehensive a control over all the dealings of their customers as did the men they were replacing.[1]

III

Up to 1875 the only large innovation in the fishing of the Shetlands was the emergence of cod fishing conducted on smacks in relatively distant waters. Even before 1800 sporadic attempts were being made to break through the limitations imposed by the use of small boats in the deep-sea fishery. One weakness of small-boat fishing was the length of time that was spent moving to and from the grounds, a necessity which, coupled with the short endurance at sea, meant that only a small proportion of the time could be spent in fishing. The range of the small boat also was limited. The use of larger vessels in the great-line fishing itself was supposedly ruled out because of the need to control accurately the speed and direction of the vessel during the operations of laying and hailing the lines. But a rôle was seen for the larger vessels in acting as 'parents' to the small boats which would continue to do the actual fishing. Thus the limits on range and endurance might be broken; boats were towed to the grounds where they made periodic returns to the parent during an extended period of fishing. The method was tried but proved abortive, because of the tendency for boats to crash as they were being towed and because of the danger of their scattering too widely in sudden storm.[2]

It was only after 1800, then, that it was discovered that, with handlines, large vessels could be used for a regular and profitable fishing. With this technique the vessel simply drifted while fishing.[3] Lines, each controlled separately by an individual fisherman, hung into the water below the vessel and the success of the method depended on fish being found in this middle water. The cod was such a fish for, although often taken on the bottom by long-line, it was also vulnerable to the hand-line.

Cod fishing from larger vessels became a profitable investment for the general merchants and curers of the Shetlands from the beginning of the second decade of the nineteenth century. As the fishing developed out of the experimental stage an annual complement of decked vessels

---

[1] *Committee on the British Fisheries, 1785,* 63; Shirreff, *General View of the Shetland Islands,* 77; Edmondston, *View of the Zetland Islands,* i, 244; *Sea Fisheries Commission, 1866,* Q. 32172; *Commission on the Truck System, 2nd Report, 1872,* 26–7.

[2] *Committee on the British Fisheries, 1785,* 25; Edmondston, *View of the Zetland Islands,* i, 251.

[3] Goodlad, *Shetland Fishing Saga,* 130.

of about 30 tons, were fitted out for operation from March to August
on a variety of near and distant grounds. They were crewed by nine to
fourteen men.[1] The cod were brought back at intervals during the
season—when distant grounds were being fished, three landings would
be made in the year's operation—to be cured by the same method of
drying as were the ling from the haaf, the curing stations being mainly
within the central area of the islands rather than on the periphery as
was necessary for the sixareen fishing.[2] The vessels were provided,
maintained and fitted out by shore-owners, and involved a substantial
long-term investment with additional annual charges which, given
reasonable catches, could be made profitable to the owners on a system
of half-shares with the crews. Each year saw its fresh arrangements,
since the men were engaged on seasonal contracts. They provided lines
and food for themselves, and were attached solely to the one vessel for
the whole annual period of fishing. The half-share which the crew took
in the proceeds of the fishing was divided between the individual members in fractions stated in the original contract. Generally, such were
the catches made at the cod fishing that the men would make more out
of it than they could from sixareen fishing.[3] It seems, indeed, to have
been an adventure that attracted young, unmarried men—perhaps in
rebellion against the traditional and somewhat unrewarding conditions
of the haaf.

Cod fishing on this model seems to have been started in 1811. For
the next six years the efforts were confined to nearby waters and no
outstanding success appears in the record, but in 1817 a profitable
fishing bank was struck near Foula and the high profits attracted more
capital to start the cod fishing on an upward swing which strengthened
through the mid- and late twenties.[4] In this period of a minor but booming fishing the sloops were provided mainly by a large number of merchants, each contributing one or two vessels.[5] For the merchants
concerned, such investments were one item in the package of their many
interests. The boom ended in the early thirties in years of relative failure
and this destroyed the interest of many of the lesser men; when recovery

[1] Seaforth Muniments SRO GD46 Sec. 13, 152/6, 'Report on the Shetland
fisheries', 1837.
[2] Smith, 'Trade in the Shetland Islands', 216; Tudor, *Orkneys and Shetlands*, 143;
Robert Cowie, *Shetland and its Inhabitants* (Edinburgh, 1874), 124.
[3] *Sea Fisheries Commission, 1866*, QQ. 31758–9.
[4] Smith, 'Trade in the Shetland Islands', 201–2; Tudor, *Orkneys and Shetlands*,
157; Seaforth Muniments, SRO GD46, Sec. 13, 152/6 'Report on the Shetland
fisheries', 1837; Andrew C. O'Dell, *Historical Geography of the Shetland Islands*
(Lerwick, 1939) 123–7.
[5] Smith, 'Trade in the Shetland Islands', 196.

began after 1835 the effort was concentrated in the hands of fewer and bigger firms which could better bear the short-term fluctuations.[1] The fishing, indeed, settled down to become a major interest of between six and ten firms, together fitting out an annual total of vessels which ranged between forty and seventy. After a short-lived drop in the number of these vessels, the number climbed speedily to a peak in 1837, but from then the numbers sank again until 1850 and rose only slowly till the late sixties. This period—when, with due allowance given to size of the vessels used, the cod fishing was probably at its highest level—was the prelude to a steep decline, probably due to the development of other fishing interests. By 1885, the 'Faroe' fishing as it had now come to be called was a small affair indeed but it continued at a low level until final obliteration about 1900.[2] When the numbers were still big in 1871, 605 men were being engaged for the cod fishing as compared with over 2,000 in the still undiminished small-boat fishing.[3]

The effort in cod fishing was, in fact, tending to expand more definitely than the figures for vessels engaged would at first indicate. At least part of the fleet was making more distant voyages and was being equipped on a correspondingly more lavish scale. In 1833 the bigger units began to venture to Faroese grounds and in the late forties (for a short period) they made trial of the Davis Strait.[4] The increasing size of the vessels and of the required investment was no doubt a factor in concentrating the business in the hands of a few big firms; and, with profits from a main growing sector of the fisheries thus concentrated, the predominance of the big firms was consolidated. In the fifties the pattern of this so-called 'Faroe' fishing, at least for the bigger vessels, came to be two trips to Faroe and a final voyage to Iceland to complete the season's work.[5] The underlying principles of financial arrangement, the procedure for hiring crew and the method of fishing itself, all remained as they had been for the first ventures in nearby waters.

Until the 1880s herring remained secondary to ling fishing. While attempts were made to sustain a regular annual effort, it was from the crews which had completed the standard seasonal fishing at a haaf station that the men were engaged. Thus herring fishing did not get

---

[1] Ibid.
[2] Goodlad, *Shetland Fishing Saga*, 158; Smith, 'Trade in the Shetland Islands', 206, 292–3.
[3] *Commission on the Truck System, 2nd Report, 1872*, 28, 30.
[4] Goodlad, *Shetland Fishing Saga*, 133–6; Smith, 'Trade in the Shetland Islands', 202.
[5] Goodlad, *Shetland Fishing Saga*, 141–4; Smith, 'Trade in the Shetland Islands', 202–3.

under way till well into August by which time the herring were past their best condition and the catch consisted largely of spent herring that brought only a low price in the market.[1] Landings were cured by pickling but, because of the low quality of the cure, the product was mainly sent to the West Indies and Ireland rather than to the rising Continental market. Boats were engaged at a considerable number of widely scattered stations which required little adaptation to be effective as curing centres, and engagements were offered on the model of the arrangements that had proved so successful in holding together the fleets on the east coast mainland of Scotland. But the rates were low and herring fishing never brought more than a small and rather uncertain addition to the main fishing earnings of the haaf.[2] Neither did it prove a venture of any great profit to the curers. Their investment consisted in part of the usual laying in of stock; but they also owned a considerable proportion of the fishing fleet, composed of half-deck but small boats acquired especially for herring fishing and hired to the crews for the seasonal use on half-catch terms.[3] One merchant firm, Hay and Ogilvy, had 100 such craft, about half the number of all specialist herring boats.[4] Landlords also retained an interest in herring fishing, although in the main they depended for their supplies on the use of sixareens which were owned by the crews. The bulk of the fleet was of this unsuitable type. Indeed, the unsuitability of many of the boats, the uncertain and small catches, the lateness and brevity of the season, all prevented the rise of this herring fishing to the point where its success might have begun to attract boats and capital from ling fishing. Catches rose to annual totals of about 50,000 barrels in the late thirties but even in these years—which represented a peak for the Shetland herring fishing before the dramatic changes of the late seventies—the average boat was landing each year less than 100 crans of inferior fish, sold at low prices.[5] Moreover, after the rising catches of the thirties the whole industry collapsed, with poor catches and falling prices. The losses made by the curers were such as to cause the withdrawal of capital and through

---

[1] Smith, 'Trade in the Shetland Islands', 206; *NSA*, XV, Shetland, 16, 21; *Sea Fisheries Commission, 1866*, Q. 32119.

[2] *NSA*, XV, Shetland, 67; *Commission on Crofters and Cottars, 1884*, Q. 18509; *Commission on the Truck System, 2nd Report, 1872*, QQ. 14109–16.

[3] W. Fordyce Clark, *The Story of Shetland* (Edinburgh, 1906), 86; Christian Ployen, *Reminiscences of a Voyage to Shetland, Orkney and Scotland* (Lerwick, 1894), 174; *NSA*, XV, Shetland, 30, 54; *Commission on the Truck System, 2nd Report, 1872*, Q. 14109–16.

[4] Smith, 'Trade in the Shetland Islands', 205–6; Ployen, *Reminiscences of a Voyage to Shetland*, 171.

[5] *NSA*, XV, Shetland, 4; Clark, *Story of Shetland*, 124.

the three succeeding decades herring fishing fell to be a tiny and sporadic affair.[1]

IV

Until 1880, therefore, the main interest of the bulk of the fishermen, year after year, continued to be the haaf. The size of the fleet rose and fell in long slow sweeps. The number of boats seems to have risen in the 1820s to 450 boats but the succeeding two decades saw decline.[2] From 1850 to 1870, production was again on the increase, although this may have been due in part to a growth in individual catches with the somewhat larger boats that were then being used. At the high level of 1870 the haaf played the same rôle for the participating families as it had done a century before. They moved to the same stations, they fished in the same manner and they supplied the raw material for the same type of curing as in the past.[3] The fishermen continued to be small-holders who would fish for a period of the year and for the rest of the time cultivate the land or fish locally in the voe for a daily supply of food. But, while the numbers engaged do not seem to have risen above the levels reached in the early nineteenth century, the total population itself had increased by about 50 per cent between 1801 and 1861. The increased manpower may have been taken up partly by the increase in the smack fishing; but the open-boat fishing, unchanged in its scale and method, was playing a smaller part in the total social life.

One major change did overtake the haaf fishing in the middle decades of the century—a shift in control from landlord to merchants which was part of a general retreat of landlords from direct involvement in trade. The landlords were content to draw their share in the profits of fishing and of trade out of increased rents rather than take the day-to-day risks of operation on their own account. The line between landlord and merchant is seldom very clear, and least of all was it so in Shetland. The landowners of the islands had been the major traders since the seventeenth century; on the other hand, many men who began as merchants purchased land both for the better pursuit of their business and for the social prestige it brought. Nevertheless, there is at least nominally a clear distinction between the established landowner who branches into trade and the merchant who makes his first money by trade and then buys land. The distinction may become muddled as

---

[1] *Commission on the Truck System, 2nd Report*, 1872, QQ. 14109–16; *Sea Fisheries Commission, 1866*, QQ. 31909, 32060, 32175.

[2] Smith, 'Trade in the Shetland Islands', 203–4.

[3] *Sea Fisheries Commission, 1866*, QQ. 31768–90; Tudor, *Orkneys and Shetlands* 129–42.

generation succeeds generation, but in the Shetland of the early nineteenth century the rise of the merchant group was so recent that most men could be firmly placed on one or other side of the line dividing landowner from merchant. And, very definitely, the second group—the merchants who might become landlords—were on the rise in the first decades of the century. Having established a position in external trade, their capital and connections were taking them into the internal trade that had for a century been carried on by landowners.[1] Shops were opened at many points for trading with the mass of the people who previously had received their goods from the landowners (who then retreated, giving up their trade business, although, at least until 1820, they held on to the profits of curing). But in the next twenty years even this branch of business began to go to the merchant-curers.[2] In fact, the reasons for the retreat are not clear. Possibly landlords found that they could live equally as well as rentiers of a trade in which the profits were maintained and even increased under the merchants. Possibly, they found it more difficult, once the original trading position had been surrendered, to block their tenantry from selling the products of fishing to the increasingly confident merchant class. Whatever the explanation, curing and the control of fishing had by 1840 also passed over to the merchants.

These merchants were, in every case, men of many interests who combined export and import with retailing of all types of goods, with purchase of produce of different types, with curing in the traditional way, and with fishing ventures of new types. Some were big men, becoming landlords of importance in addition to engrossing a large proportion of general trade. As far as fishing was concerned, in spite of their swift development of the smack fishing, it was by their control of the haaf that they most widely affected the population. Here lay a large part of their varied trading interests and investments. In 1867 four-fifths of the product of the haaf was sold by nineteen merchants; and, within the group, two firms (each with over 200 fishermen on its books) employed between them 24 per cent of fishermen and the top nine firms were handling the business of 66 per cent.[3] This trade was notable not only for its concentration in the hands of a few firms but also for the way in which the individual merchants carved out particular areas of control within which all the trading transactions of the individual customer were managed by the one firm, just as completely as they had been in the days

[1] Smith, 'Trade in the Shetland Islands', 225-6.
[2] Ibid., 225-6, 228.
[3] *Commission on the Truck System, 2nd Report, 1872*, 25.

of the uninhibited supremacy of the lairds.[1] As in the earlier days, a fisherman would have to sell his fish to a given individual who would also supply him with his daily necessities—with his boat and gear and with whatever accommodation in credit he needed. The merchant seemed as much in control of the price he paid for fish and charged for his goods as had been the laird.

It is true that, under merchant control, the tenure of land was less overtly used as a means of directing the tenant to fishing under given conditions although the merchant might well, by receiving a tack or even purchasing land, take over the landlord's rights to the labour of the fishermen. Or the merchant might lease the fishing rights, including the right to the products of the tenants' fishing. Or he might simply acquire the ground for a store. There was no reason, then, why merchants should not have continued to use the fishing tenure as an instrument of compulsion. Yet, on the whole, formal compulsion to fish and to sell to the laird or to his tacksmen tended to be less and less used. In 1870 compulsory power was asserted by landlord, tenant or lessee over some tenants on some estates; but more numerous and influential were the landlords and merchants who claimed that tenants and fishermen were not thus formally bound.

In fact an apparatus of controls, more subtle but equally persuasive, was being exercised by the new merchant class. Such compulsion to fish was largely unnecessary in an agrarian system where the cost of holding on to land from which subsistence was derived was met out of income which could only be found at the fishing. Increase of population meant more families seeking land, and this prevented any increase in the size of holdings to the level where they could provide a full and independent livelihood. There was some reserve of land waiting to be brought into cultivation and, indeed, the area of cultivation was somewhat expanded under population pressure; but the growth of the improved land was less than proportionate to the rise in the population seeking land and the tendency was for holdings to decrease in size. In addition, by deliberate policy, landlords split the available land into smallholdings of standard size—holdings which would provide some subsistence but which would yet be too small to divert people from fishing. This subdivision does not seem to have caused the degree of fragmentation that was common in the western Highlands, and holdings

---

[1] The argument of the following pages is based on the *Truck Report on Shetland* (1872). In that report, and more particularly in its Minutes of Evidence, the economic relationships of the main groups in Shetland, as they stood in the late sixties, are displayed in the most intricate and telling detail.

of four to six acres were common; with the extension of potato acreage they would provide something close to a complete subsistence. But the need for money, to pay the rent and to purchase the increased amount of goods that were imported rather than locally manufactured, drove most of the men to play some part in the fishing.

The product of this fishing was channelled into the hands of the dominant merchant group by a number of devices. Sometimes the lease of land for a store would state that no other merchant was to acquire the land (and with it the opportunity of trading with the tenantry of the estate). And where the merchant became a true and fairly large land-owner, his power to keep out intruders was so much the greater. Where exclusive sway was not guaranteed by formal regulation, other means might be used to give the merchant power over the fishing population. The form of the contract for the sale of fish was such as to bind the fisherman or crew to one merchant for long periods: the normal method of selling was by an exclusive engagement for the season; to sell at all, the crew had to agree to deliver the complete supply to the one merchant. Such arrangements, we have seen, were not unusual in the fishings of the Scottish mainland and they were not inconsistent with keen bargaining before the seasonal contract was agreed. But in Shetland there were two features that diminished the worth of the bargain to the fisherman: firstly, there was very often only one man with whom he could agree and the alternative to acceptance of terms was not to fish at all; and, secondly, there was no stated price in the initial agreement. The price was retro-spective, decided only when all deliveries had been made, and indeed well past the end of the fishing season in October or November—by which time the merchant had taken delivery and the fisherman had no power of bargaining. The only restraint on the merchant in deciding on a price was the desire to retain goodwill for the following season; and where he had no local competitor this goodwill might be of slight com-mercial importance. In some districts there was, however, apparently the possibility of switching custom from one merchant to another and an implicit competition tended to keep the price of 'green' fish near the price of the cured equivalent. But, even when there was choice of dealer, fishermen seldom switched from one to the other: inertia seems to have played a considerable part in maintaining the exclusive hold of each merchant over his customers. On the whole, the merchants' power of deciding price seems to have been moderately applied. As in the days when the lairds had been the purchasers, the price of 'green' fish was adjusted to that of cured fish, rising and falling as the market rose and fell. The different merchants paid a uniform price to their customers. A curer, therefore, had a guaranteed supply of fish obtained on terms which gave him a not unreasonable margin but which secured him against any possibility of loss.

The contract which channelled the results of a crew's fishing to one curer inevitably involved other forms of trading in which the Shetland curers were no less interested than in their dealings in fish. Fishermen had to wait for settlement till long after the end of the season, and they needed advances to keep them going in meal, trade goods and fishing equipment. Since the proceeds of the season's fishing came directly into his hands, the merchant with the contract for the fish had the security against which to make a loan and it was always he who was the source of credit. This was granted in the shape of the direct provision of goods and, apart from the interest which might be charged, it became a means of expanding the general business of the curer. The fisherman became completely dependent on the curer for his supply of trade goods; and the prices he would be charged, like the price of the fish he had to sell, were determined by simple edict, retrospectively, when the general settlement was drawn up in the autumn of the year.

When the time came to renew the boat or replace the lines, the crew would particularly need help; still drawn from among the associates of an agricultural township, they might hire their boats for a fee which was deducted from the gross takings before the residue was shared equally among the members; but more usually they would buy the boat, in order to become the full owners, usually at the cost of running a debt for a period of years. Again the inevitable source of the boat, or of the finance for the boat, was the curer. Similarly, lines and hooks would be acquired from the common store, again generally on credit. Thus the capital transactions by which crews acquired the means of fishing were embodied in the general account, as items along with the debits and credits for general supplies, meal, agricultural produce and fish. The loan and the hiring arrangement for the boat were, of course, transactions with the whole group, which became collectively responsible for repayment. But the individual's share of the group transaction was kept separate, to be embodied in his part of the particular account with all its personal items.

Thus the account between the fisherman and the curer-merchant embraced many items and covered most of the commercial transactions of any sort in which the two parties would engage. Sometimes the tenant would emerge clear of debt at the end of the season and a cash payment would be made to him; but this was inevitably small in comparison with the value of all the transactions made in his name: in the main this was a system whereby the merchant supplied the particular individual with the necessities of life and capital equipment, and received in direct return his produce of fish, knitted goods and agricultural items. Sometimes, indeed, no cash would change hands at all, the tenant ending the day in debt which would be carried over to the following year. One type of transaction had been removed from this general account: in this period the payment of rent was a separate matter. But even this separation was more apparent than real. Merchants pursued their businesses on sufferance of the lairds and they were sometimes held responsible for the payment of rents out of the credits in the tenants' accounts.

Clearly this was a system which gave almost tyrannical power to the merchants. They could manipulate the prices almost as they wished and they could dictate a flow of credit on which the tenants' livelihood utterly depended. The interest charged was concealed in a price system which consisted of mere ciphers in a general account. Yet there is no indication of feelings of rebellion on the part of the tenantry. Given the choice of masters they would often continue without question to channel all their dealings through the one store and one source of credit. The acceptance of the nearly permanent control of one man or firm had its roots in the feeling of need for some greater protection than could be found in an open market, for someone who would supply all needs, come what might, than in any direct compulsion.

CHAPTER VIII

Herring at its Peak: the Older Communities, 1884–1914

I

FROM THE early fifties to the early eighties herring fishing grew, generally in near boom conditions which were never interrupted for long. As always with herring fishing there were large annual fluctuations but there could be no doubt that the trend of production was upwards, particularly in the seventies and early eighties, the increase being sustained at first by the use of more boats and then by improvement of gear and catching capacity, by developing new areas for fishing and by lengthening the season. The incomes of fishermen, always vulnerable to fluctuation, tended to increase and much more steadily wealth accumulated in the shape of the equipment of the trade. Curers' profits were probably even more widely variable than fishermen's earnings, but the conditions making for loss never lasted for long and there were always new aspirants to fill the gaps left by those who failed. Yet there was weakness behind the optimism and the prosperity. Curers had advanced the expansion of the industry by running deeply into debt; and they had become accustomed to promise to the fishermen rates per cran which could only be met, without further borrowing, if the price of cured herring—always fragile—continued at a very high level. Even a smallish drop in this price, once the engagements for a season had been made, was certain to cause losses for most of the curers. It had happened even in the years of prosperity and there were always prophets of disaster to be heard, but price recovery had always come quickly and the lending banks showed themselves willing to support most of the curers through times of difficulty. The danger was that as productive capacity grew the market would collapse for a period of years, that curers would be unable to set the losses of one year against the greater profits of the next, and that the banks would be forced to allow many of them to fail.

In 1884 just such a crisis of over-production came. At the beginning of the season every available boat was under engagement and curers had agreed to pay bounties of over £60 to each crew, in addition to the basic rate of 20s. per cran (prices which allowed no profit to curers unless the price of Crown Brand Full herrings held above the high level of 35s. per barrel).[1] Curers had been fairly clear of debt at the end of 1883 but by the time they had made their preparations for the new season they were deeply in debt again, mainly to the banks.[2] Early in the year, however, there were ominous signs from the Continent, where stocks were high at the beginning of the year, and there was growing competition from Scandinavia.[3]

The first real trouble came with a Lewis fishing which was very plentiful but of inferior quality. The main market for Lewis fish of the early summer was in Germany, among the middle-class consumers who ate high-quality herring along with the expensive spring vegetables; lower grade fish were difficult to sell at this time of year because they were mainly eaten with potatoes that did not come in quantity till later. Thus a comparatively small quantity of Lewis fish could have a disproportionate effect on price. Speculative confidence, which alone kept back stocks from the market, was beginning to evaporate.[4] The next trouble came from Shetland. The enormous fishings of that area added to the impression that masses of unsaleable fish were accumulating.[5] Finally the main east coast fishing added to the perplexity in two ways: firstly, it was plentiful; secondly, it was composed in large proportion of immature fish which brought a much lower price in the market than fulls, and curers had to pay as much for such fish as they did for the higher quality.[6] Prices fell steeply and curers were left with stocks of inferior fish, with heavy commitments to the fishermen, with a burden of debt and with no possible retrieving action. The greater part of the season's catch was sold at a curer's loss which in many cases was as high as £1 per barrel.

These losses were not at first perceived to be other than a short-term aberration and the banks continued to support the curers, carrying the debts over to the next year and even pumping in fresh credit.[7] As a result, while for the next two years lower rates were agreed per cran, there was little diminution in the fishing effort, although the drop in incomes and the poorish prospects together with the lack of bounties all slowed down investment by fishermen. Boat-building was immediately

---

[1] *Fish Trades Gazette*, 12 July 1884.   [2] Ibid., 13 July 1884.
[3] Ibid., 30 May 1884.   [4] Ibid.
[5] Ibid., 23 May 1885, 3 July 1886; *Peterhead Sentinel*, 30 Jan. 1888.
[6] *Fish Trades Gazette*, 9 July 1887; 25 Feb., 10 Mar., 7 July 1888.
[7] FB Rep., *1887*, p. xxxviii; *1894*, 172; *Daily Free Press*, 23 Dec. 1887.

halted and the buying of new nets slowed down.[1] The large catch that resulted in 1885 and 1886 kept prices low, although the high quality of the 1886 herring helped to improve the return; by 1887 it was beginning to be understood that the capacity for catching herring in Scotland was indeed too great for the immediate market and that prices would remain low until either the catch was diminished or some long-term change occurred in the market.[2] Thus the banks, now deeming that some of the curers' debt was irretrievable, accepted the loss of assets and withdrew support.[3] Failures among curers followed; coopers were thrown out of work; fishermens' incomes were depressed seriously by unwillingness to engage boats.[4] More and more the tendency was for fishermen to hire fewer extra hands and to leave boats on the beach through the fishing season. The fishing effort slowed down. Another change that was started was for curers to buy by auction rather than at a price agreed beforehand and fixed for the season as a whole: in this way, they felt, they would more easily keep their buying price at levels that were safe in terms of the ruling price of cured herring. The fishermen were generally against such a change; and, indeed, in the smaller and more remote stations there tended to be too few buyers to create a fair market for the fishermen. But, slowly, after many backslidings and at different pace in different centres—in Peterhead first, then in the other large centres, then in the small—the curers forced on the new auction system.[5]

II

It took herring fishing nearly ten years to begin to show signs of recovery from the crisis of 1884. From 1886 to 1893, by the index of a moving five-year average, catches remained below those of the early eighties. Price as well as output stayed at low levels and the small prospective income from herring prevented any great new effort, either in terms of the number of boats prepared for the fishing or of investment in a renewal or extension of equipment. Then in 1893 came the first upward jerk due to a high yield from a still restricted fleet. With prices also showing signs of improvement, an upward spiral started, and the average cure of the years 1894 to 1898 was 6 per cent above that of the boom years of 1881 to 1885. This upward trend was to continue although there were signs of a downturn just before 1914. The greatest total of herring cured in any period of five consecutive years

[1] *Fish Trades Gazette*, 4, 11, July 1885.
[2] Ibid., 12 Sept. 1886, 7 Jan. 1887.
[3] *Peterhead Sentinel,* 23 Sept., 8 Nov. 1887; 31 Jan. 1888.
[4] *Fish Trades Gazette*, 9 July 1887; 25 Feb., 10 Mar., 7 July 1888.
[5] FB Rep., *1887*, p. xxxviii; *1894*, 172; *Daily Free Press*, 23 Dec. 1887.

was achieved from 1907 to 1911. By then the average yearly total was 76 per cent above the average year of the mid-nineties and 59 per cent above that of the booming early eighties[1] (Figs. 15 and 16).

In this period, outside the trawl ports of Aberdeen and Granton, herring fishing had come to dominate any other type. In 1913 the value of all herring landed in Scotland amounted to £2,087,754. The east coast fishermen were producing at least three-quarters of this and, since they also produced catches valued at £763,256 in East Anglia, their earnings from herring cannot have been less than £2,250,000. Against this, all other types of fish (apart from landings from the trawlers) gave a value of £270,000.[2] The story of the life of the three-quarters of the east coast fishermen who still lived in the traditional, scattered and fairly small communities was now the story of the herring.

The boom in herring was touched off by high catches from 1893 onwards, taken with equipment which was in much the same state as it had stood for about ten years; and it continued because high catches were later accompanied by rising prices. Only a small proportion of the total catch went to the home market where there was little chance of expansion for fresh herring as the supplies flooded in. Inevitably, the increased output was cured and sent to the Continent; and these supplies, moving along the well-established channels to the ports of Germany, Poland and Russia, were sold at generally increasing prices. The tendency during this expansion was for a greater proportion of exports to go to the eastern end of the market area, to Russia.[3] Indeed, some of the cure which was landed at German ports was destined to be sent east, into a broad area of Russia now served by the extended railway system.

After 1900 good sales and a slight improvement in prices were more and more turned to profit which, in turn, initiated an investment boom resulting in an extension of catching power. The first improvement, occurring before 1900, was the widespread purchase of boats equipped with steam capstans by which the crew were enabled to handle eighty nets at a time, thus increasing their catching power by as much as 20 per cent over the previous best.[4] Then after 1900 came the much bigger step to the use of the steam-powered drifter. Experiments with steam vessels for drift-net fishing began as early as the seventies; and no very difficult technical problems needed to be solved for steamers to bring back, reliably, a considerably higher daily yield than the sailers.[5] Since steam trawling was in rapid expansion from the

[1] FB Reps.   [2] FB Rep., *1913*, App. B, no. II.   [3] FBR, 62/318.
[4] FB Reps, *1894*, 156, 160; *1895*, 172.   [5] *Fishing News*, 5 May 1914.

Fraserburgh

Price paid for uncured herring (per cran)

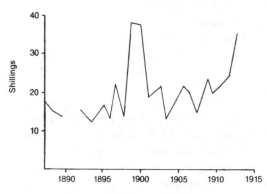

Source: FB, Annual Reports.

Price per barrel of cured herring
(Crown Brand Fulls)

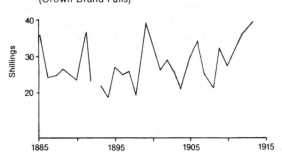

Source: FB, Annual Reports.

early eighties, it may seem strange that no big move was made to
develop steam herring fishing until after 1900. The explanation probably lies in the relation of yield to the costs of operàtion. The financial
advantage of changing over to the steamer was, at the very least,
dubious when prices and yields stood as they did in the nineties. Not
only was the initial expenditure on the purchase of a new vessel itself
very heavy—at least three times as much as for a sailing boat—but also
a steamship was burdened by heavy operating costs which were pretty
well fixed whatever the catch. Only when gross yields rose above a certain point did the steam drifter make a profit even on the narrow defi-

nition of profit as simply covering the working expenses of the season. There were conditions of price and catch, then, under which the sailing vessel might catch less but be more profitable, or where it might make a profit while the steamer was making a loss. When physical yields and (less certainly) prices rose in the late 1890s the line was crossed at which the steamship appeared more profitable; and the advantage became even greater in the years 1905 to 1907 when gross earnings on both types of boat reached unprecedented heights.[1] The result was something near to a mania for steam drifters, starting with relatively modest investments in the new type from 1900 to 1904 and then reaching a climax in the great building programme of the next three years. From 1908 till 1914 investment tailed off, but even by 1908 there existed a fleet of steam drifters which could catch more than had all the sailing boats of 1900 and which by then dominated the catching of herring, although there was still in operation a big fleet of sailers employing more men than did the steamers.[2]

The framework which held the expanding herring fishing was fashioned by the curers. The pre-condition for successful herring fishing, at any time of year and on almost any part of the coast, continued to be the gathering of curing stock and working staff at given stations. The herring caught around the Clyde estuary were largely sold fresh, but elsewhere curing by pickling remained by far the most important means of converting to a saleable product the landings which came in great surges (as they had always done). The technique of curing had changed little, even while the catching power of the fleet was being multiplied several times over; and the herring catch had to be swiftly gutted and laid in salt if it was not to be wasted.[3] Profitable fishing depended therefore on the curers making preparations and having a labour force ready at hand. Important changes were made, however, in the way in which curers acquired the catches. One of the consequences of the crisis of 1884, as we have seen, had been an unwillingness by curers to continue to guarantee prices to the fishermen before the season started. At the main centres they could be sure that on any given day a large number of boats would be making landings, and it best protected their interests to offer the price that seemed suitable from day to day in accordance with the factors of general supply and price of cured herring. Sales by auction now seemed to suit them best except in the very remote stations. Fishermen had resisted the change to auction-selling,

---

[1] See below, pp. 160-2.
[2] *Committee on the North Sea Fishing Industry, 1914*, App. 19, p. 222.
[3] Duthie, *The Art of Fishcuring*, 5-20.

but by the early nineties they had been compelled to go over to the new system in all but a few outlying areas.[1] They did not come to be any less dependent on the curers—it was indeed the curers' strategy of investment, however it might originate, that determined where and how long the herring fishing proceeded.

The main centre of interest continued to be the 'Great Summer Fishing' off the east coast itself, but two important changes were made in this now-traditional herring fishing. The season was lengthened by an early (June) start; and curing was largely concentrated in the districts of Wick, Fraserburgh, Peterhead and Aberdeen which by 1913 accounted for 89 per cent of the east coast catch. If Wick is left out, 71 per cent of the catch had come to be made on the small stretch of coast of the north-east corner. Within each district there was now great concentration on the main ports and all but three out of the 229 curing stations of their districts were in the ports of Wick, Fraserburgh, Peterhead and Aberdeen.[2] This centralization was the result of the use of large boats which, even before the time of the steam drifter, gave an overwhelming advantage to those places which in the seventies acquired deep-water harbours of some size. But proximity to the best fishing grounds also played its part. Buckie had a good fishing harbour, yet only a few boats used it for herring fishing. Wick was still an important port but was, increasingly, surpassed by the centres to the south.

The interests of many of the curers from the North East spread far beyond their home area. They were involved in most of the developing fishings in other parts of the kingdom and by their investment in remote stations they provided the opportunity to the fishermen of virtually a year-long herring fishing. Their interest in the west coast was of long standing. Since the early 1840s they had been organizing a string of simply equipped yards along the edges of the Long Island lochs as well as in the more highly developed centre of Stornoway. For many years the predominant interest in the area was the early summer fishing at which at one time a fleet of over 1,000 boats was collected. But increasingly it was the winter fishing, with a single base at Stornoway, which formed the main investment. By 1902 some 180 boats, nearly all from the east coast, were collecting for this winter fishing and by 1913 the figure was 158, almost all being steam drifters which were better fitted to the winter conditions. In the meantime the early summer fishing

[1] *Daily Free Press*, 23 Dec. 1887, 8 Sept. 1894; FB Rep. *1887*, p. xxxviii; *1894*, 172.
[2] FB Rep., *1913*, App. A, No. II.

had so declined as to occupy little over 100 boats, among which the incomers at the summer fishing would leave before the end of May.[1] The period of the Shetland herring fishing largely coincided with that of the home east coast summer fishing. It had originally been developed in the eighties by mainland curers and was then followed by a large force of mainland boats; it continued as a major if distant interest for the curers, particularly those of the North East. Shetland in these summer months attracted boats from the east coast, from England and from the west coast. In general, total landings were somewhat below those of the east coast mainland but individual catches were high and the mainland always sent a large contingent to fish in Shetland waters rather than at home. Indeed, considerable numbers of east coast crews were tempted to pass the summer fishing in Shetland rather than on their own coast.

Shetland might provide an alternative for the summer months, but the biggest innovation was the autumn fishing in East Anglia. As we have noted this area had begun to attract boats from Fife as early as the sixties but it was not until the 1890s that crews from other parts began to interest themselves in it. By 1900 the Buckie district was sending about half its fleet, which then still consisted of sailing vessels, to participate in the main East Anglian fishing, from October into December.[2] Some of the districts at that time still sent only small contingents, but in 1913, when 1,163 boats made the voyage, nearly all the main herring fishing districts had a heavy interest in East Anglia; both the catching and the curing at Yarmouth and at Lowestoft were not only a main interest for the Scots but also were locally dominated by them.[3] In particular, nearly all steam drifters were taken to one or other of these ports. For the individual boat the yields of this fishing were very much on a par with those made on the native east coast.

Thus, at least for the steam drifters, the two main possible occasions of high earning were the home (or Shetland) summer fishing and the East Anglian autumn fishing, with almost equal chances in each. These, providing a fishing lasting almost continuously from June to December, were the firm and invariable basis of the fishing routine of the steam drifter: none could afford to miss out on either of these voyages. The west coast winter fishing, on the other hand, was less likely to yield big gains and only a fraction even of the steamers would take part in it.

The development of these various seasonal fishings, through the

---

[1] Ibid., *1902*, 225; *1912*, 221.   [2] Ibid., *1900*, 235; App. A, No. II.
[3] Ibid., *1913*, p. vi.

marketing opportunities largely provided by the Scottish and east coast curers, gave crews the chance of spending much of the year in pursuit of the herring. In the nineties, such fishing was not evidently more profitable to the fishermen of the east coast than the fishing for cod and haddock on which they still spent much of their time; and annual earnings often showed the returns from white fishing ahead of those from herring.[1] Yet herring fishing had always been tempting because of the possibility of high gain in a short period; and with equipment becoming ever more expensive it was advantageous, even necessary, to use it as much as possible. Thus, in the gathering boom of the nineties there were already signs that line-fishing was playing a smaller part in the lives and livelihoods of east coast fishermen.[2] After 1900, indeed, the high gains to be made at the summer fishing and at East Anglia tended to draw more crews into herring fishing, but it was the steam drifter that finally tied a large proportion of the east coast men to herring fishing alone. Great-line fishing was technically feasible from the steam vessel, but fishermen apparently considered that their best chances of covering the heavy operating costs was by herring fishing. Most who acquired steam drifters, then, drove them as hard as possible in the main herring fishing seasons and turned to line-fishing only occasionally, and only when no main herring fishing was possible.

Steadily, therefore, fishermen outside the trawl ports were drawn into complete dependence on herring fishing and into an annual routine of moving to the main fishings of the west coast and of East Anglia. And it was in the North East, where of course the biggest fishing was now firmly centred, that the herring ruled most strongly. Herring had for some time been caught by north-east fishermen in almost any of the fishing areas around Scotland and off eastern England but their catching of white fish had been carried on largely from local bases. When, therefore, we find their local landings of cod and of haddock falling to negligible levels it is a sign of their relinquishing almost completely—except for a small amount of great-line fishing in distant waters—an interest in white fishing that goes back to the first records of their activities. This was the area in which the steam drifter took firmest hold. By 1914 only a small majority of the men on the line of coast between Peterhead and Nairn were engaged on steam drifters; but they were catching very much more than half the fish accountable to the people of the area.[3] Thus, not only had they become in general highly

[1] Ibid., *1893*, 152; *1894*, 156.
[2] *Daily Free Press*, 4 Jan. 1904; FB Rep., *1907*, 261.
[3] FB Rep., *1914*, App. B, No. I; *Committee on North Sea Fishing Industry, 1914*, App. 19, p. 222.

dependent on the herring but they were involved in a new set of social relationships arising out of the use of steam drifters. On the east coast as a whole by 1914, only in the Montrose district were there landings of white fish which were apparently large enough to be a significant factor in fishing incomes.[1]

III

The relatively high cost of the steam drifter and the lack of the opportunity to buy at second-hand brought a decisive change in relations among fishermen and between the fishing and the outside financial communities. Ownership by fishermen of the vessels they used remained an effective feature but the circle of owners was very much narrowed as landsmen took a stronger control, entering as full owners as well as extensive creditors. About 1900 a drifter cost about £1,500 as compared with the £700 to be paid for a sailing boat of the latest and largest type; but when the steamers began to be built more stoutly the price rose until, by 1912, the very minimum was £2,700, while £3,200 might be asked for a steel vessel.[2] To this sum had to be added £600–£800 for the 200 nets necessary to a vessel which was to be used almost continuously for the several herring fishings of the year. Such sums were beyond the immediate means of the small partnerships that traditionally had financed the building of new vessels. Yet while the typical group that would find the money was still the two- or three- or occasionally four-man partnership, the result of the deal left the fishermen (while nominally independent) to work under a much closer control from the shore. A partnership which disposed of a sailing boat to buy a drifter would have at the most, apart from savings out of earnings, £500 or £600 together with the nets which they had used with the earlier vessel—a long way from the cost that faced them in buying the new unit. Aid might be sought in the form of a loan for which the vessel would be mortgaged, and the first likely source was the bank. On such a loan, the conventional limit was one-third of the value of the purchase, the loan being assigned to the group as a whole rather than the individual members. This might well still leave a large gap and other lenders would then be tapped—such as merchants, curers, rope- and sail-makers and, more particularly, fish salesmen. If enough was raised by borrowing to make up the full price, the fishermen-partners would then be sole owners of the boat and nets, but with heavy debt

[1] FB Rep., *1914*, App. B, No. I.
[2] *Committee on the North Sea Fishing Industry, 1914*, QQ. 1115, 4365. The discussion of the problems of ownership and share in the new drifter fleet is based on the voluminous detail of this report.

charges to be met. However, the only way they often had of raising the full sum was to allow landsmen to take full owning shares in the boats; the fishermen would still keep to themselves the major interest but perhaps one-third of the profits would go to the landsmen-owners. Here, as well as in the provision of loan-capital, fish salesmen were the main participators.

In fact, the fish salesmen were coming to exercise a strong and pervasive control in addition to drawing profit and interest from the operations of the fleet. The profession of fish salesman had been created when the old system of engagements to curers had collapsed in the crisis of the late eighties. Fishermen now required agents to act for them in the daily auctions at which their catches were sold and normally a crew would make a long-run arrangement with a salesman who would dispose of their entire landings at a commission of 5 per cent. The group that rose to implement this new system of selling in the nineties was to become the wealthiest and most influential within an increasingly complicated industry, almost entirely displacing curers as the main source of funds for the various needs of the industry.

They came to command enterprises of considerable size and of varied function. Since boats moved to different parts of the coast it was necessary to have agents or branches in many different ports and, out of the simplest relationship implied in the selling of the catch, the salesmen came to handle the profitable business of supplying gear and stores. Some developed the manufacture of ropes and sails. By extending loans and by a more direct participation in the ownership of boats, they came also to a control of the whole business side of a boat's operations. In particular, they played an important part in the raising of the loans for initial purchase. When a partnership of fishermen was looking for accommodation at the bank or for help from others, a direct approach would be made by salesmen who might have some money of their own in the venture. In effect they acted as sureties for the fishermen, a position they could adopt because of their close interest in the practical management of the affairs of the small enterprise that resulted from each new boat that was built and equipped.

Control by salesmen was certainly solidified by their grip on the funds needed by the fishermen. It was questionable whether the provision of a loan to potential owners carried with it an obligation by the borrowers to use the services of the salesman-lender as manager or entitled the salesman to take the profitable business of supply as well as the inevitable right of selling the catch; but, whatever the formal obligation, there is no doubt that owners tended to hand over all the dealings connected with the running of their boats to the men who had supplied directly, or had obtained access to, the funds for purchase. Control was not only intricate and penetrative but also widespread. Thus the large business of the port of Fraserburgh, at which was landed 40 per cent of the total catch of the summer fishing, was apparently in the hands of three main firms; and these same firms of course, had agencies spread over all the fishing ports.[1] In Aberdeen, Meff Brothers had the management of 100 drifters and direct shares in seventy of them.[2]

As the tentacles of the shore firms—as owners, creditors and managers—spread through the fleet, a new structure of ownership emerged. For the first time in herring fishing a substantial direct share in fishing boats was held by non-fishermen and, moreover, by enterprises whose interests spread over many separate fishing units. The particular ownership structures of the drifters ranged from complete ownership by fishermen to the situation where landsmen owned the boats and put in hired men as crews. Thus in 1911 some 403 out of 783 steam drifters were owned wholly by fishermen; at the other end of the scale 115 were owned wholly by landsmen.[3] The boats in which fishermen had no direct share were heavily concentrated in the Aberdeen, Peterhead and Wick districts; but in Fraserburgh, Banff, Buckie and Anstruther, where most of the remaining drifters were to be found, by far the greater share remained in the hands of the fishermen. But this conveys an unduly favourable impression of the independence of these fishermen who, even when they were nominally complete owners, were without doubt heavily in debt. Thus 556 of the 624 vessels which were owned in whole or in part by fishermen were mortgaged, and some loans had been raised other than by mortgage.[4] Banks and salesmen each held 21 per cent of these mortgages, being followed at a distance by merchants and fish-curers as the other main groups. The management of the boats, too, was largely out of the hands of the fishermen who did, however, retain their traditional right to decide where and when to fish.

Equally important were the changes in the distribution of ownership within the fishing community. There were six or seven fishermen-members of a drifter's crew, but seldom did more than three of them have shares in the boat, and frequently there would only be two owners within the crew. In all, the number of owners had shrunk to be a minority among the men who actually worked the steamers. In part this was due to the fact that the men who provided the capital in loans

[1] Ibid., Q. 735.   [2] Ibid., App. 1, p. 189.   [3] Ibid., App. 19, p. 221.
[4] Ibid., App. 18, p. 220.

liked to have a compact group to deal with; but no doubt another factor was that, even with the greatest possible subdivision of shares within the crew, the minimum share needed for ownership was far above the capacity of the majority of fishermen.

Nets, which represented a distinct and not inconsiderable investment in the fishing venture, were provided in part by the owners but in part also by men for whom this was their sole financial contribution; but, if ownership of nets was rather more widely spread than that of the shares in the boats, there still remained a group which had no property in the venture at all. Hired men were of course no new thing in herring fishing; but in the past they had come from outside the fishing community, as a means whereby capital, in the shape of boats and the gear in which shares were widely and evenly distributed across the fishing community, could be made to bear more profit. On the steam drifters, however, many of the hired men were fishermen indigenous to the communities in which the owners also lived, and a strong and new line of social division was thus being marked across the community of equals. In Buckie, Fraserburgh and Peterhead, each with a large complement of drifters, between one-half and one-third of the fishermen engaged on steam drifters had shares in them.[1] The proportion of fishermen who owned nets was somewhat higher, being as much as 95 per cent of all fishermen in the Buckie district, and 50 and 40 per cent respectively in the Fraserburgh and Peterhead districts. But there were residues of fishermen with no property at all, ranging from 60 per cent of all fishermen in Peterhead to 5 per cent in Buckie. And it is likely that in Aberdeen and Wick, where the proportion of shore-owned drifters was high, the proportion of fishermen without property would also be high. These figures relate only to those sections of the fishing community which were permanently engaged on steam drifters and they show the divisive effects of that type of operation. Yet even in 1911 there were still many sailing boats in full use, and with them more old-fashioned schemes of ownership prevailed; consequently each district had elements of the old egalitarian diffusion of property mixed in with the new and more hierarchical order. Thus, every man of those engaged in sailing vessels would have some nets, and in Buckie and Fraserburgh 48 and 82 per cent respectively had some share in the boats. This type of arrangement still provided the basis of livelihood and social position for the big majority in Fraserburgh while in Peterhead slightly less than half and in Buckie 41 per cent still worked sailing boats by 1911.

These divisions were possibly less rigid than they seem on a static

[1] Ibid., App. 19, 222.

enumeration. Ownership of nets might be a step to higher things, earning profits out of which came a full share in the boat, while some of the unpropertied were doubtless sons and younger brothers who might eventually inherit or have assigned to them a share. But even by 1911, when the main drifter fleet was scarcely five years old, there had been no time for such diffusion to take effect; and the war and the post-war years brought an end to new building, a restriction in profits and a general scarcity of capital—and therefore did not allow the less privileged to improve their position. The lines of division as drawn in 1911 were to be very difficult to cross in the ensuing years.

Clearly there were emerging big differences in the wealth of different sections of the fishing community. The owner of, say, a third of a drifter and of a quarter of its nets had a worth of over £1,000, while the man next door might have no property at all. The differences in earning capacity, on the other hand, are less easy to trace. While those with capital invested in the process of fishing were likely to earn more than those with none, the relative incomes of the different groups might vary a great deal from year to year; a profitable or a poor fishing had varying but specific effects on owners and on hired men.

The basic determinant of incomes was the arrangement for sharing proceeds among the owners (of boat or of nets) and the mere deckhands who owned none of the equipment. After 1900 the great cost of upkeep and replacement of the boat, and the increased value of its gear created the need for a new system of sharing, a system different from the one which had long prevailed on sailing boats (where the proportion of earnings assigned to the boat was too small for the new circumstances). Somewhat different schemes of sharing were adopted in the different districts but the general principles are sufficiently clear to allow generalized description.[1] Normally each fishing season, such as the summer herring fishing, the East Anglian venture and the winter fishing in the west, was considered as a separate enterprise with its own accounting and share-out: but, while all boats participated in the summer and East Anglian fishings and while there were other possible ventures, the profit outcome of the year as a whole depended largely on the combined accounts for these two 'voyages'.

It was a marked and occasionally disastrous feature of operation by the steam drifter that heavy expenses had to be met which bore no relation to the size or value of the catch; within fairly narrow limits of fluctuation, these were fixed and had to be borne whether the gross earnings were great or small. As compared with the scale of earnings

[1] Ibid., App. I, 185-92; FB Rep., *1919*, 26-8.

that had been feasible before the coming of the steam drifter, these out-goings were very high indeed. The main items charged against gross earnings before any division was made were the coal bill and wages for the engineer (and possibly for the fireman and cook) who were paid on a fixed wage basis and were not normally men of the fishing com-munity. Commission on sales, another big item, did of course fluctuate with the size of the catch; but for the most part working expenses had no relation to the gross earnings of the vessel. Generally speaking, for a boat engaged in herring fishing for nine months of the year expenses were between £600 and £700 with a possible tendency to slight increase just before 1914.[1]

The different items of expense might be charged against gross earn-ings in various ways. Always a large part was deducted from the gross total before any share was allocated to the various groups, but there were one or two items (such as payments to the cook and fireman, cutch for nets, and insurance) which might either be taken out of the gross or charged against a particular group. In fact it did not make much difference to the outcome how these particular weights were adjusted. The sum that remained when expenses had been deducted was divided into thirds, one to go to the owners, one to the nets and one to the crew. The crew's share was further divided into six—that is, it was allocated to individual owners of boats and nets as well as to hired men. The third share which went specifically to owners did not, in its entirety, represent spendable income; out of it there had to come upkeep, amounting to about £230 on the full year, and an allowance for de-preciation reckoned at £150 to £180. Out of the net's share had to be made an allowance for replacement and repair, estimated normally at over £300.[2]

Steam drifters in their short pre-war history had a chequered record of earning power, and on the whole tended to increase the fluctuations in the yearly earnings of the individual members of the fishing com-munity as well as to produce a more definite social stratification. Not only did they make fishermen more completely dependent on the most uncertain type of fishing but also the inflexible character of many of the outlays increased the uncertainty of the earnings which remained to be disbursed. The first two years of the twentieth century, when the fleet of steam drifters was still very small, saw big catches giving impetus to investment in boats of the new type. But by 1903 and 1904 earnings had

---

[1] For example, FB Rep., *1903*, p. xxxix, Tables I–II, and similarly in the reports up to 1914.

[2] *Committee on the North Sea Fishing Industry, 1914*, App. I, pp. 188–9.

dropped away, and it is doubtful whether in these years many of the steam-boats in fact paid their way. In 1905, however, there started a succession of three prime years with high earnings and rapid increase in the steam fleet. The boom ended in 1908, the worst year that the drifters had ever known; and earnings continued to be very moderate till the period 1911–13 which saw prosperity ascending—and in the last full pre-war year, 1913, the returns of the previous best year (1906) were nearly equalled.[1]

The meaning of prosperity and relative failure can be seen by examining specific years. The best year of all was possibly 1906.[2] The median value of the average gross earnings at the different stations then stood at £2,321, representing earnings, net of expenses, of about £1,700. At this level owners would comfortably meet depreciation, insurance, interest and upkeep charges and still have a surplus of £200, together with their share of crew earnings. One man's share—such as would go as sole earning to those without nets or a share in the boat—would be £94 for the whole year. Individual boats made earnings of up to £4,000 which meant £130 in the year for each member of the crew, together with a surplus for the owners of the boat of £450; an owner of part of the boat and of some nets who was an active member of crew might then take a profit of about £390, above all charges. But these were the very top earnings in the very best of years. Even in that year, there were also some boats which did not produce more than £50 for their crew members and failed to do more than cover the expenses of the owners.

The results, taking one year with another, can best be seen by looking at one district, that of Peterhead, and examining the results for all the pre-war years.[3] Of the seven years from 1902 to 1908 three were outstandingly good in terms of the whole pre-war experience of the steam drifter, but two were poor indeed. In this period, in one of the better years, earnings per man (excluding the return that went additionally to owners of boats and nets) on a boat making average returns would have been slightly upwards of £90 for the year; that is, a man without a share in boat or nets would be making the equivalent of a moderately well-paid industrial wage-earner. But in the two bad years earnings fell away to £37 and in the very worst year to £27, poor remuneration by any standard. Similarly, owners of boats and nets must have taken more than enough to meet the charges specific to them—insurance, interest, upkeep and depreciation—in five out of the seven years, but would have failed to do so in the worst two years. For the remainder of the

[1] FB Rep., *1903*, pp. xvii–xix, and similarly in the annual reports up to 1914.
[2] Ibid., *1906*, Tables I–III.   [3] Ibid., *1905–14*.

pre-war period the information is less complete but it is clear that the three closing years were nearly up to the best, and none fell to the worst, levels of the previous period. Personal earnings, then, would always be of £50 or more, with three occasions on which they probably topped £90; and in all years there would be a profit for owners of boats, which probably increased to £200 to £300 per boat in the best three years. The individual owner of boat and nets would be making close on £300 clear of all expenses in these good years; indeed, in the total run of the years from 1902 to 1913, such men would have been earning at this level, on the average boat, in six out of twelve years, some compensation for the two occasions on which they probably lost on the year's operations.

IV

The final stages in the intensification of herring fishing, first with large sailing boats and then with steam drifters, brought considerable shifts of the fishing population. The east coast had communities scattered over its entire length—amounting to over 100 in number. In the different phases of the development of line fishing and of the initial rise of herring fishing, when mobility was increasing and equipment improving, these communities had grown very much in step, but from the eighties the uniformity of their experience was shattered. Even while the national output and the value of equipment was growing rapidly, some fishing communities were in positive decline: complex forces were eroding their position. Thus the growth of large-scale fishing business in Aberdeen and Granton not only tended to attract men from the surrounding small places to take part in trawling but created a market structure so superior in its facility that even line-boats began to find it expedient to work out of the larger centres. The men who moved from the Kincardineshire villages to the growing Aberdeen suburb of Torry did not all come to work on trawlers. At the same time developments within herring fishing had a similar if less forceful effect. Heavy boats of over 50-foot keel came to form the main herring fishing fleet and were difficult to maintain at many of the smaller places which traditionally had depended on herring fishing as a main source of livelihood. Fishermen in some of the small creeks were forced to maintain their boats at some distance from home. Where there was a good nearby harbour, the men of villages without shelter for the boats might well be able to carry on as herring fishermen owning their boats. But where they were at a greater distance from any haven they might well find the ownership and maintenance of the largest type of boat impossibly burdensome. Some left their old homes, to give up fishing or to pursue it in a larger centre; others continued to use small boats for local inshore fishings combined with a turn as wage-earners at a major centre.[1] By 1900, then, a growing movement was under way, drawing the population out of the smaller places, particularly on the middle section of coast between Peterhead and the Tay and along the east coast of Caithness.[2] Where there was no local centre to draw this migrating population, the decline of the small creeks might mean a decline in the district as a whole; but in other cases it simply meant movement towards a growing point nearby and the fishing district figures would not be unduly affected.

These forces, indeed, hit the different districts somewhat selectively. The competition of Aberdeen was felt most acutely from just south of Peterhead as far as Montrose. This was an area which had always possessed many small villages with negligible harbour accommodation —the type of place which found it difficult in the new technological age of fishing. Stonehaven, it is true, prospered; both in boats and in numbers of fishermen it was much greater in 1914 than it had been in 1886.[3] But the other fishing communities which lay so thick both to the north and to the south were in steep decline. Some of the villages lost all their fishing population, others dwindled into tiny residues of their former selves. In 1886 there had been twenty-five settlements between Peterhead and Montrose; by 1914 five had disappeared entirely and nine had dwindled to contain less than twenty fishermen. Some of the residual communities were left with no first-class boats at all, their inhabitants' only livelihood being in small-scale line fishing from yawls or in hiring themselves as hands on the bigger boats owned by the men of other parts. The communities which had harbours capable of sheltering sailing vessels would often retain some boats, still able in the conditions after 1900 to make reasonable incomes out of herring fishing. They continued the old-standing practice of fitting out for the summer herring fishing, often moving to one of the active centres farther north to pursue it. These vessels were owned on the old terms, with most of the crew—apart from hired hands who were in any case outsiders—holding shares in the nets and vessel. Both for herring fishing and for line fishing they retained an old social system of equality and general sharing under still moderately profitable terms.

[1] *Aberdeen Free Press*, 6 Jan. 1903; *S.C. on Sea Fisheries, 1893-4*, QQ. 6490-2, 6515-16.
[2] FB Reps, *1886*, App. D, Table VII; *1900*, App. A, No. II.
[3] FB Reps, *1886*, App. D, Table VII; *1914*, App. A, No. II.

Fife had a string of communities each of considerable size. They all had harbours which, however, were mostly tidal and of little use for the large boats. But the Union Harbour of Anstruther served the neighbouring ports and allowed the Fife men to use their large boats locally. Fife fishermen had always been willing to look to distant places for their fishing opportunities and kept themselves well to the fore when new boats and equipment were in view. Some crews acquired steam drifters which were used in the summer fishing farther north and in East Anglia; and, quite commonly, in their own district for great-line fishing. Such vessels, as everywhere, created divisions between the owning fishermen and those who provided only their labour. But there was still in this area a considerable complement of sailing vessels—some with motor power—in which the old concepts of sharing prevailed.[1] The towns of the East Neuk from Crail to St. Monance, held their population and sustained a considerable local fishing. But farther west along the Fife coast the small places were in decline, and even the major centre of Buckhaven lost most of its fishing populace.

Along the southern shore of the Forth, only Eyemouth, Port Seton and Newhaven survived as substantial fishing communities with their own fishing equipment. Newhaven used small vessels for local, mainly herring, fishings in the Forth, but the others lived mostly by herring fishing at a distance.[2] There had been a limited investment in steam drifters in Eyemouth but sailing vessels, all operated almost entirely in distant ports, generally provided the main means of livelihood.

On the long coast north of Inverness, there was general decline which affected most sharply the small places of the southern part of this stretch, although it extended north as far as the major centre of Wick. In Caithness the herring fishing had once been conducted partly from a chain of tiny coves, some of which did not even have artificial shelters for the boats; the fishermen who lived there were mainly part-time, combining their fishing with farming, and sometimes fitting out their boats only for two months in the summer. The coming of big boats meant that local herring fishing was no longer possible. Some crews would then move into Lybster or Wick for the summer fishing and lay up their boats for the rest of the year.[3] By the eighties, however, such spasmodic activity was no longer sufficient to maintain a first-class boat of latest design. Many places lost their first-class boats and some were

[1] *Committee on North Sea Fishing Industry, 1914*, Minority Report, 173–4; App. 19(a), 221.
[2] *R.C. on Trawling, 1884–5* QQ. 4772, 6586.
[3] *Departmental Committee on the Sea Fisheries of Sutherland and Caithness*, P.P. 1905, *XIII*, 7; QQ. 766, 837, 841, 1106, 1338, 1354, 1669.

left with no fishing population at all, although some continued on a reduced scale with a local inshore line-fishing in small boats, eked out by earnings made as hired hands on bigger boats owned elsewhere. Three or four centres managed to hold on, using sailing vessels in herring fishings at a distance combined with local line-fishings. Wick and Helmsdale invested in steam drifters but by far the larger part of their labour force was still employed on sailing boats, for which they had the advantage of a local summer fishing for herring.[1]

The fishermen of the line of coast from Peterhead northwards to Fraserburgh and then westwards to Nairn saved their position in the herring boom by large investments in the means of catching herring, particularly in steam drifters. Nor was it only the fishermen of the larger centres where active fishing was now concentrated who thus improved their equipment; in the smaller villages investment was just as active. Some of the village owners were able to bring their boats home when they were not in use; but there were those who did not even have this convenience and who, with the decline of line fishing after 1890, had come to use their home villages simply as places of residence. The main centres of Fraserburgh, Peterhead, Buckie and Lossiemouth grew rapidly and in 1914 they contained 38 per cent of the whole fishing population of the North East.[2] But the middle grade of villages also grew, if rather more slowly, so that by 1914 some 44 per cent of the fishing population were in villages each containing between 200 and 400 fishermen, which could now be regarded as the typical fishing community of the region. In such a unit the great bulk of the population was still devoted to the one occupation and overwhelmingly the main source of livelihood for this increasing population of the region was herring. Most of the herring were caught in steam drifters but the labour force was almost equally divided between those involved on steamers and on sailers. Thus there was intricate inter-weaving between the social patterns, associated on the one hand with the sailing boat in which nearly all local participants would have shares and on the other with the steam drifters which opened up big divisions and brought separation between families even within a close-packed community. Buckie was the district which had invested most in steam drifters, and there the population was overwhelmingly involved in their working. At the other end of the scale was the Fraserburgh district in which the bulk of the people still had some share in sailing boats.[3]

[1] Committee on North Sea Fishing Industry, 1914, App. 19(b), 222-3.
[2] FB Rep., 1914, App. A, No. II.
[3] Committee on North Sea Fishing Industry, 1914, App. 19(b), 222-3.

CHAPTER IX

The New Community: Trawling, 1880–1914

I

IN 1880 the main fishing activity at Aberdeen was its part in a summer fishing in which the port occupied an important but by no means dominant position. Few of the boats which worked out of Aberdeen for this seasonal fishing belonged either to it or to the nearby fishing communities, for the resident fishing fleet consisted only of a handful of sailing boats owned by the men of the tiny settlements of Torry and Footdee. The year-round activities conducted from the port by boats of local ownership amounted to a minor inshore line fishing, the landings from which made up a tiny proportion of the total catch similarly landed in the many creeks and harbours all up and down the east coast of Scotland. In addition, occasional stranger vessels such as trawlers from England might call to land the results of their fishing in the vicinity. Yet by 1890 over 250,000 cwt. of fish (other than herring) were being landed at the port and within ten years average annual landings had quadrupled to over one million cwt.; by 1914 this had doubled again.[1] Even by the early nineties Aberdeen was certainly the greatest white fishing port in Scotland, accounting for some 20 per cent of the catch (apart from herring); by 1913 this proportion was 70 per cent.[2]

In part, the growth of business reflected the appearance of a fleet of steam trawlers, owned locally and operated almost solely from the home port. In the operation of trawling a bag-net is dragged over the sea-bed by the motion of the fishing vessel. The descriptive term generally applied in the nineteenth century to the technique—'beam trawling'—referred as well to the wooden beam which was used to hold open the mouth of the net. It was a form of fishing long-established in England and after 1840 had taken a modern and highly organized form, though still conducted in sailing vessels, with the rise of the east coast ports of Hull and Grimsby. English trawlers often worked northwards along the Scottish coasts but most of the Scottish fishermen were not only resolutely opposed to the incursions of the English boats but also unwilling

---

[1] FB Rep., *1901*, App. B, No. I; *History of a Great Industry* (Dundee, 1903), 18.
[2] FB Reps, *1891*, App. C, No. I; *1913*, App. B, No. II.

to develop similar methods of fishing on their own account.[1] One or two of the east coast settlements, it is true, did equip their sailing boats for some seasonal trawling but this was the reviled activity of a minority of fishermen in only one or two centres.[2] Then in 1882 came the first experiments in steam trawling from Aberdeen; and the success of the one boat which was fitted out in the first season and the two in the second detonated the first trawling boom, in which the number of trawlers, all steam-powered, increased—with only two exceptions—year by year until 1913.[3] By 1893 Aberdeen had thirty-eight steam trawlers, by 1900 it had 100 and by 1913 as many as 218.[4]

Trawlers were specialist vessels operating intensively throughout the year and using very effective forms of fishing gear. Thus, before long, white fishing in Scotland was dominated by the trawler and, after an initial period during which Granton equalled Aberdeen in the size of its fleet, it was the trawlers operating from the northern port which accounted for the great bulk of the catch. Yet the business of the port of Aberdeen was much more than an annual operation by its own fleet of trawlers; rather, the growth of that fleet was itself in part due to the autonomous development of port facilities there. Trawlers first began to make landings at Aberdeen because they made good catches in the near vicinity, as is clear from the fact that the Aberdeen boats immediately established a highly profitable routine in which very nearby grounds, in Aberdeen Bay and farther north, provided the opportunity of daily return to port with full holds.[5] Fuel costs were low and little time had to be wasted in steaming. Nothing more was needed to bring outside boats to a port with such good local opportunity and to stimulate local effort in the equipment of a fleet to make full use of its situation. The initial spurt in landings soon started a chain of developments which absorbed the results of fishing in distant grounds, which led more crews—not only of trawlers—to make the port their home, which stimulated further investment in trawling and which created the conditions whereby the port would act as a focus for the boats of many districts and nationalities. Aberdeen, in fact, became a market centre pulling magnetically upon a fishing fleet of diverse origins and stimulating local investment in the means of fishing both in near and in distant waters.

When landings began to arrive in increased volume, the first step was

---

[1] R.C. on Trawling, 1884–5, QQ. 963–8.   [2] Ibid., QQ. 1317–78.
[3] S.C. on Sea Fisheries, 1893–4, Q. 6450; R.C. on Trawling, 1884–5, Q. 1518.
[4] FB Reps, 1893, App. A, No. 1; 1900, App. A, No. I; 1913, App. A, No. I.
[5] R.C. on Trawling, 1884–5, QQ. 1521, 1619, 1726, 1738, 1839, 2175; S.C. on Sea Fisheries, 1893–4, Q. 6450.

to establish a daily auction at which any vessel making landings would have the opportunity of selling to assembled buyers.[1] Hitherto the fish had either been hawked direct to the consumer or had been sold to fishmongers who acted as retailers. Some of the local fishmongers, too, were establishing a country trade by which they distributed, partly as wholesalers, the fish purchased by casual and personal contact.[2] By the other main system of sale, found often in the haddock fishing, the buyer would engage boats and would take the whole seasonal product at a fixed price; but such arrangements were obviously unsuited to attract the custom of stranger boats on which for a considerable period most of the business of Aberdeen was to be based.[3] It was the daily auction, attended by the general body of local buyers, which gave the chance of good prices both to regular and to occasional customers. Auction selling also created an opportunity for specialist intermediaries. A fish salesman would be engaged to aid the fishing crew's contact with the market; he acted as agent in selling, keeping in steady business connection with particular boats and their owners, so that a close relationship, with considerable ramifications, developed between the owner or owners of each boat and the salesmen who customarily operated on their behalf. The salesmen, in fact, taking a fixed proportion of vastly increased sales, became a wealthy group of great influence and widely diverse interests. Out of their accumulation of capital came the funds for development in various sectors. By 1913, in the web which interlocked interests in selling of fish, in trawl-owning, in ice manufacture, in coal merchanting, in insurance and in fish processing, it was hard to discern the point from which the control had started.[4] But it was the fish salesmen who stood most at the centre, and in most cases their business as fish salesmen had pre-dated the growth of their other interests. It was always a small group—much fewer in number, for example, than the curers who mostly remained separate and who were generally each in a much smaller way of business with fewer outside interests. In 1900, a mere fifteen firms handled the whole selling business of Aberdeen, which at the same time contained nearly 100 curers.[5] In the first generation, as the port began its growth, they were men of somewhat diverse origin although generally, before beginning to specialize their business, they had had some connection with dealing in fish. William Meff had been a fish and game dealer but by 1884

---

[1] *R.C. on Trawling, 1884–5*, Q. 1963.   [2] Ibid., Q. 1968.
[3] Ibid., QQ. 1983, 2023, 2046, 2221.
[4] *Post Office, Aberdeen Directory, 1913–14.*
[5] *Post Office, Aberdeen Directory, 1900–1.*

he could be described as a 'licensed fish salesman and auctioneer' as well as still a 'game merchant'; in due course his sons were to form the influential partnership of Meff Brothers, with a business centred on the selling of fish but also with many other interests.[1] Thomas Walker had been a curer before he became a fish salesman as well as a trawl-owner.[2] There were outsiders, too, who were attracted by the evident growth of business in Aberdeen. Peter Johnstone had been fifteen years in business before he came to Aberdeen from Birmingham.[3] He travelled around the coasts, to Scarborough, Whitby, North Shields and Grimsby as a fish buyer. Arriving in Aberdeen in 1881, by 1884 he considered it to be his home. Besides acting as salesman he bought fish, mainly from the line boats, for packing for home and foreign markets.

The regularity of supply which resulted from the daily auction meshed with increased demand. In particular, Aberdeen became notable for its heavy demand for haddocks. Behind this lay an increasing capacity within the city for the smoking of the fish. The curing of haddocks by various local traditional methods had, throughout the nineteenth century, been the mainspring of much of the winter fishing, at which small boats worked inshore from a string of bases. In the seventies, some of this dispersed catch was still processed in primitive kilns attached to the cottages of the fishermen, with the product being either directly retailed or sold to fishmongers in Aberdeen or the lesser towns; but an increasing quantity was being purchased as wet fish for smoking in larger units.[4] The curers in turn sold most of their output to the Glasgow region. When Aberdeen became a large supply centre for fish, curers could be sure of some purchases at prices fluctuating according to the scale of landings but generally within expected and calculable limits, and a curing industry which had been scattered up and down the coast began to concentrate there. Increasing numbers of buyers came to the Aberdeen quays and prices were correspondingly sustained. The switch to Aberdeen was cumulative; as the crews which had worked from the small creeks saw the strong and regular market in Aberdeen so they began either to run for the port or even to take up residence in the growing centre; and the struggling remnants of an industry depending on its interlocked ports found it difficult to survive in the small nearby centres. Aberdeen tended to suck in the processing section of the industry, as well as the fishermen of the small settlements for some

---

[1] *R.C. on Trawling, 1884–5*, Q. 1931.
[2] *Post Office, Aberdeen Directory, 1880–1.*
[3] *R.C. on Trawling, 1884–5*, QQ. 1997–2001.
[4] *S.C. on Sea Fisheries, 1893–4*, Q. 6490.

distance north and south—north almost as far as Peterhead and south beyond Stonehaven. Supplies from the line boats rose along with the increasingly dominant supply from the trawlers, local and stranger.[1]

The high price at which haddocks sold at Aberdeen, the main attraction making for the increase of the business of the port, was made possible by the proliferation of small firms engaged in the curing of the fish. Of some importance too was another form of curing—the drying of cod, by methods in which the fish was laid on racks and subjected to artificial draught and heat rather than left to the old sun-drying process on the open beach. Between them these two forms of curing accounted for the greater part of the demand in the first decade of growth; but the bulk of the catch was of haddocks and by far the greater part of it went to the smoking kiln.[2] In 1890 the sale of fresh fish in the markets of the south was still of relatively small account. It is true that in the mid-eighties the North British Railway had started to run a special fish train daily, to leave Aberdeen in the early afternoon and reach London in the early hours of the following morning—so that fish landed and sold in Aberdeen of a morning would be again sold at Billingsgate within twenty-four hours. But until 1890 the service was little used and often the consignments were insufficient to fill two trucks.[3] The earliest growth of the Aberdeen trade was very firmly based on local curing and on sales either nearby or in the south-west of Scotland. A large proportion of the catch was of haddocks, of which six-sevenths would be cured; with cod, which might also be cured, bulking large in the residue, it was not necessary to find a large market for fresh fish.

By 1892, then, Aberdeen was firmly established as a main port. The previous seven or eight years had seen an almost invariable steep annual increase in landings. The local trawl fleet consisted in 1892 of thirty-eight vessels; line boats had flocked to the port; trawlers from other parts used it regularly; and casual landings, not least from foreign vessels, made a fair contribution to the total. Much of the fishing was still done in nearby waters so that even steam trawlers were not long absent from port; in fact, the advantages of an established system of marketing and strong local demand had been added to earlier attraction of prolific local grounds. In the nineties the pattern of fishing began to change so that in trawling and in great-line fishing—mainly conducted from steamers—Aberdeen began to draw from the produce of grounds in which it had little remaining advantage of proximity. One

---

[1] Ibid., QQ. 6490–2; *Daily Free Press*, 1 Jan. 1892.

[2] *S.C. on Sea Fisheries, 1893–4*, Q. 6514.

[3] *Fish Trades Gazette*, 3 May 1890.

reason was the prohibition of trawling within the three-mile limit, another was the closing of the Moray Firth to British (although not to foreign) trawlers:[1] the solution was to go farther afield. At different times of the year the Aberdeen-based boats would go east and south-east to North Sea grounds (Smith's Knoll and Dogger Bank), north to the waters of Shetland and Orkney, north-west to the Minch and Atlantic, and by 1893 they were venturing to the Faroes and to Ice-land.[2] The North Sea remained the chief fishing area but supplies were now coming from areas where Aberdeen often had no advantage of location. For the more northerly grounds, it is true, there was some obvious gain in using the northerly port. Fish landed in Aberdeen from them could reach London more speedily than fish from the same areas landed in Hull or Grimsby, since transit by land—by rail—was substi-tuted over part of the distance for transit by sea. Transport costs by rail were, of course, higher; but with high quality fish the advantage of speed—and of high price—might counterbalance the disadvantage in cost. However, when fishing was pushed to the northern waters, Aber-deen faced the prospect of having to compete with other trawling ports, and the costs of operation inevitably increased. More powerful and more costly vessels had to be built; fuel bills were higher; time was lost in steaming at the cost of time for fishing.

In these more demanding conditions the growth of the port—and its fleet—continued unchecked. Landings increased four-fold between 1890 and 1900.[3] More and more in this period Aberdeen had to hold its business by its services as a market. A major improvement was effected by the opening in 1889 of a municipally owned fish-market which pro-vided covered accommodation into which the boats could directly unload. With the railway station nearby, a special siding allowed the fish to be loaded direct on to the trucks that would go south.[4] The curing stations to which much of the catch continued to be taken were some-what more widely scattered, so that the physical transportation of fresh and cured fish followed complicated patterns within a fairly small area. Much of the handling was in small quantities and there was necessarily much duplication of journey when the fish was moved to the curing yard and then back to the station. Yet, with all its defects, the Aberdeen fish-market was widely regarded as one of the attractions of the port for foot-loose vessels; and Aberdeen's building initiative here was

[1] *Daily Free Press*, 2 Jan. 1893, 1 Jan. 1894; *S.C. on Sea Fisheries, 1893–4*, Q. 6354.

[2] *S.C. on Sea Fisheries, 1893–4*, QQ. 6382, 6408, 6464, 6597; *Daily Free Press*, 1 Jan. 1894, 4 Jan. 1899.

[3] FB Rep., *1901*, App. B, No. I.   [4] *Fish Trades Gazette*, 25 May 1889.

imitated by Dundee in its effort to achieve greater status as a fishing port.[1] Some of the difficulties of the marketing of fish at Aberdeen arose not so much from the concept and planning of the fish market as from the increasing volume of business, which soon produced overcrowding and consequent serious and costly delays.[2] The successive extensions which ultimately created the greatest fish-market in the country, running to one-third of a mile in length, generally lagged behind the declared need for quay-space adjacent to the market; but they surely indicate both the success of the market system and the great desire to use what facility there was.

The sheer volume of landings meant an increase in the supply of fish of types which could not be cured and for which the main market was in England; the flat fish, in particular, found their best sale there. The efficiency of the railway connection, then, became more important. In 1889 the Caledonian Railway followed the example of the North British in providing a fast daily service by special fish trains from Aberdeen to London and as total landings increased in the nineties the consignments, mainly of flat fish, created a worthwhile traffic.[3] The cost of moving the fish was 4s. 8d. per cwt., a substantial addition to the price paid in the Aberdeen fish-market,[4] but a cost which could be met because Aberdeen's fish rated as high quality when sold in London. It had the edge over Hull and Grimsby in respect of quality because the trawlers on the average stayed at sea no more than six days and their catches did not suffer the extra day's steaming to the English ports. The London trade thus became important for the trawlers making landings at Aberdeen.

Yet haddocks remained the most important element in the catch. In the late nineties they accounted for between 40 and 50 per cent of the value of the catch and, while the proportion tended to diminish slightly after 1900, this remained the most important single type of fish to be landed.[5] There was little sale for them in England and the greater part were still cured for distribution in south-west Scotland.[6] It was therefore the growth of the curing trade that continued to be the main strength of Aberdeen as a market and to attract an ever-increasing number of daily arrivals. The price of haddock at least partly justified Aberdeen's boast that it offered the best prices in the kingdom and there was an appreciable margin between the prices of haddocks in Aberdeen and

---

[1] Ibid., 22 Jan. 1889.   [2] *Aberdeen Free Press*, 6 Jan. 1903, 4 Jan. 1904.
[3] *Fish Trades Gazette*, 3 May 1890.
[4] Ibid., 1 Mar. 1890; FB Rep., *1894*, 141, 151.
[5] FB Reps., App. B, No. I.   [6] *Daily Free Press*, 4 Jan. 1899.

elsewhere.[1] The third main component in the catch was cod and codling. Again this was mainly used for curing, in this case by drying. The port had its facilities for the curing of cod as well as for the smoking of haddock, but there was not the same evident advantageous price margin as compared with other centres.[2] In any case, the cod came mainly from Iceland waters which were little used by the Scottish trawlers. This trade, then, remained a somewhat specialized section in which foreign trawlers, particularly of German origin, landed catches made beyond the normal range of the local vessels.[3] Nevertheless the cod made a call upon port facilities, contributed to the overcrowding which threatened the efficiency of the whole fleet and provided the base for the common provision of services which helped to increase the lead of Aberdeen as the main trawling port.

One of these services was the provision of ice, increasingly important as the average absence from port became longer and as the volume of the traffic in fresh fish increased. By the late nineties a gap had emerged between need for ice and local supply, and the shortage was made up by import from Norway. The addition of a third ice factory fully met the demand[4] and from then on the port could meet its needs from the product of its own factories.

The growth of market and supply facilities thus helped to keep Aberdeen's business growing when she had lost the advantage of local low-cost trawling grounds. But at the same time, the local trawlers were finding inevitable difficulties as they were driven to widen the sphere of action of an industry which had been founded on local resource. It became necessary to build larger and more expensive boats even while operating costs increased because of longer steaming times. The change began in the early nineties and was much helped by the high prices which ruled until 1895. Even so, with the average landings per day's fishing scarcely above those made at much less cost in the eighties, the building of the new local fleet was not easy. Incoming vessels were still in the majority, and it was they which mainly kept the volume of landings rising. In fact by 1894 the Aberdeen fleet had only increased by two boats over the thirty-eight of 1890. But the composition of the fleet had changed, with a section now comprising large vessels fitted out for trawling at distances of up to 300 miles; some Aberdeen boats, however, were still designed only for near-water fishing no more than eighty miles out.[5] By 1895 prices had fallen severely and momentarily

[1] Ibid., 2 Jan. 1893, 4 Jan. 1900.   [2] Ibid.
[3] FB Reps, *1920*, 88; *1922*, 24; *1923*, 38.
[4] *Daily Free Press*, 4 Jan. 1899, 4 Jan. 1901.   [5] Ibid., 1 Jan. 1894.

the limits of expansion seemed to have been reached. But the halt was short-lived and by 1897 rising prices combined with steady and considerable annual increases in landings-per-arrival to produce a boom in which the local fleet was yearly increased by rapid steps.[1] At last locally registered boats had come to account for the greater part of the local landings.

One reason for the steep increase in the weight of fish landed by each arrival at the port was the adoption of the otter trawl, by which the mouth of the net was held open by boards set on a vertical plane rather than by the horizontal beam.[2] It provided an increase in catch of about 20 per cent for a tiny capital outlay and general adoption was almost instantaneous. With heavy catches coming from the distant grounds and with the use of the new apparatus, catches soared. By 1903 the average weight landed by each arrival was nearly twice what it had been only eight years earlier, a result achieved by a steady upward swing which had not yet ended. The number of trawlers registered in Aberdeen more than doubled within three years and by 1904 stood at 178 as compared with forty less than eight years before. By 1904, however, prices were falling; and, with catches not now increasing so dramatically, the boom slowed down. Yet, while prices did not begin to rise out of the trough till the immediate pre-war years, there was still profit in trawling and, with some fluctuations, nearly every year saw its additions to the trawlers in the local fleet.

II

The growth of trawling and of great-line fishing, and the concentration of small-line fishing in Aberdeen, led to a great increase of employment. Between 1881 and 1911, in fact, Aberdeen was the most rapidly growing city in Scotland; and fishing, with its ancillary branches, was the most rapidly growing of its industries. The bulk of those employed in connection with fishing remained ashore in plants engaged in the processing of fish, in the repair and building of boats and in the general supply of the fishing fleet. In 1905 some 9,200 persons other than fishermen were employed in connection with the industry.[3] Of these 3,000 may be set to the account of the highly seasonal herring fishing in which the yards worked only for a few weeks in the year. The remainder, however, were connected with an industry which produced steadily through the year, and most of them must have been full-time workers. They had no particular connection with a defined fishing community,

---

[1] Ibid., 2 Jan. 1893.   [2] Ibid., 1 Jan. 1898.
[3] *Aberdeen Free Press*, 4 Jan. 1905.

being hired for the most part on the general labour market of the city. Yet these calculations take no account of the many industries and activities which, though not dealing in fishing materials, experienced an increase in demand because of fishing; the labour force permanently dependent on fishing was considerably and incalculably larger than the number of those engaged in the processing of fish or in the supply of a recognizable fishing requirement.

While the fishermen, then, formed only a minority among those employed in connection with fishing, their number also showed at first an abrupt—and then a fairly rapid and continuing—increase from year to year. By 1886, the originally trivial numbers of the Torry and Footdee settlements had increased to 440 and by 1913 the city had 3,023 fishermen, there having been two periods of particularly rapid increase in reaching the final figure.[1] Such growth was unprecedented in the history of Scottish fishing communities, not only because of its rate and eventual scale but also because of the nature of the process. Fishing populations in the past had always grown organically out of established communities by natural increase or by a very slow immigration of people who were soon absorbed in the manners and relationships of an existing settlement; even where there was more abrupt movement of people it usually took the form of the transplanting and rooting in a new environment of a group which did not have to change its ways. But in Aberdeen there were no such organic connections with the past, and even when people of old fishing tradition enrolled with the new fleet they had to suffer a complete break with old relationships and ways of working.

In part the trawling fleet squeezed a labour force out of the older fishing communities by its very effectiveness as a competitor, and Aberdeen drew on the old fishing populations of the coast all the way from Peterhead in the north to Johnshaven in the south. One simple cause was that the rewards of a deckhand on a trawler were greater and more certain than those of a man with a share in a small-line boat at one of the smaller creeks. Thus, when the increase in the size and expense of herring boats left many of the men of the outer ports with no direct means of sharing in the herring fishing except as hired hands, a lower grade of fishermen had appeared at many of the small places, having nothing but small boats which they might use in local line-fishing, to be combined with the uncertain prospects of engagement as hired hands. Trawling offered a better and more certain livelihood than these men could find with their own boats and many deserted their native villages to engage

[1] FB Reps, *1886*, App. D, Table VII; *1913*, App. A, No. II.

at fixed wages on the trawlers.[1] The growth of Aberdeen as a market centre tended in another way to break down the economy of the small village which had been based on inshore fishing for haddocks for sale on contract to curers. The larger port with its better prices and easier handling led the crews of small-line sailing boats at first to run to the port and then to take up permanent residence there. For a period a considerable fleet of small-line sailing boats operated successfully from Aberdeen alongside the more powerful trawlers and the great-liners (which were mainly steamers operating in distant waters). But, by the late nineties, small-lining both in Aberdeen and along the coast had begun to shrink.[2] To the north of Peterhead the decline of small-line fishing was a sign, and perhaps a consequence, of attraction into an increasingly specialized herring fishing; but southwards along the coast at least as far as Johnshaven, except in the case of Stonehaven, decline in the local fishings meant the virtual collapse of the local community. The small villages of this line of coast showed sharp decreases in population between 1886 and 1913 and in part these were undoubtedly a sign of the haemorrhage of the line-fishing population into trawling.[3] It might take the form of the wholesale transfer of the whole community to Aberdeen where the members could take jobs in trawling and its ancillaries. In 1900 the village of Oldcastle was reported to be on the point of such wholesale removal.[4] More commonly, however, there was a partial and continuing flow, family by family. At the same time the decline of small-line fishing in Aberdeen itself increased the supply of trawlermen, for the skills of these traditional fishermen were sought by the trawl-owners. It is not clear how frequently the fishermen who were deserting their homes in the coastal villages in fact moved to fishing jobs in the larger centre. The increase in the trawling labour force roughly matches the probable volume of migration out of the villages but, since some of the new trawlermen were landsmen with no previous fishing experience, it is probable that some were completely deserting the calling of the fishermen. Engineers and firemen, for example— forming a third of the total crew—were not normally drawn from the general fishing community.[5]

Even if the new labour force engaged in trawling consisted of men of an older fishing tradition, the change to trawling meant for them a

[1] *Aberdeen Free Press*, 6 Jan. 1903; *S.C. on Sea Fisheries, 1893–4*, QQ. 6490–2, 6505–6.

[2] *Daily Free Press*, 4 Jan. 1900; *Aberdeen Free Press*, 8 Jan. 1904.

[3] FB Reps, *1886*, App. D, Table VII; *1913*, App. A, No. II.

[4] *Daily Free Press*, 4 Jan. 1900.

[5] *Committee on North Sea Fishing Industry, 1914*, App. 1, p. 190.

deep change in their way of life. Most of them became wage-earners, paid largely on a time basis, with a very small addition related to the size of the catch. At first these wages were the subject of individual bargaining between particular owners and their crews but by the 1890s crew members were acting to obtain a general wage arrangement. The engineers were organized in a trade union but it was more difficult to achieve a permanent association for the ordinary deckhands.[1] They were early in acting collectively, however, for in 1903 there was the threat of a strike which resulted in the reference of the dispute to a conciliation board and an award which accepted the deckhands' case.[2] Engagements, however, were purely for the one voyage, usually of less than a week; and it was customary at the end of a voyage simply for the skipper to ask his men if they wished, as individuals, to sail with him for the succeeding one. Sometimes it was difficult to secure the return to the ship, on time, even of the men who had agreed to a further voyage. In 1904 an attempt was made to enforce prompt return by making payment of back-wages dependent on punctuality, but the arrangement proved unworkable.[3]

Not only had the general run of trawlmen become wage-earners with only a casual interest in the boats on which they sailed, but also among the crew there arose a much sharper differentiation than had ever been known among line fishermen. Skipper and mate stood very much apart from the rest of the crew, with a much more permanent involvement in the operation of one boat, with a critical interest in the size of the catch and with a standard of earnings well above that of the deckhands. Their reward was a share in the profits of each particular voyage. The skill of the skipper, indeed, was the key to the success even of boats in company ownership, and sharing of gross earnings was a device to ensure the steady services of a good cadre of skippers. The structure of earnings going to these diverse parties is shown by the agreement of 1913.[4] In the operations of each vessel, expenses—including wages—were deducted from the gross earnings to give the balance on which the remuneration of skipper and mate depended. This balance was divided into fourteen shares of which the master took $1\frac{3}{8}$ and the mate $1\frac{1}{8}$. Their remuneration was clearly subject to wide fluctuation but might be very high. The first engineer was paid at the rate of 8s. 4d. for each day at sea, and the second engineer at 6s. 6d. Fishermen were paid at rates which differed with the size of the vessel. In the smaller type

---

[1] Ibid.
[2] Ibid., 6 Jan. 1906; also *Daily Free Press*, 4 Jan. 1901.
[3] *Aberdeen Free Press*, 6 Jan. 1903.   [4] *Fishing News*, 28 Nov. 1913.

(up to 97 feet) they would get 5s. per day plus 4d. per £ of the net balance, on the larger 5s. plus 3d. per £. Different rates were paid for time in port.

III

Some skippers obtained an owner's share as the price of their services and a very few eventually became the owners of the vessels in which they sailed; but in the main the trawling fleet was owned by landsmen.[1] The first efforts at fitting out trawlers were made by men of established business positions, usually of a type that had some connection with fishing; the earliest examples in Aberdeen being Pyper, a merchant, and Walker, a fish-curer.[2] The investment of the high profits of the first few years extended the interests of the innovators and by 1892 Pyper is found with interests in ten trawlers and Walker in seven.[3] Johnstone and Sherrett formed another block of interests, Johnstone being an immigrant fish salesman. The senior partner in Meff Brothers, William Meff, was a fishmonger who became a fish salesman and then broadened his interests into ownership and agency for drifters as well as trawlers.[4]

In the first period of growth of the trawling industry in Aberdeen a comparatively few owners, making generally high profits, extended their individual fleets.[5] But, even as the few firms grew, capital was continually being sucked in, often in small amounts, from other spheres. Pyper's interest in the individual boats of his group of vessels was in many cases as an agent for units which were 'privately owned'. Sometimes a man or partnership would be the apparent owner of a number of vessels, but in each of these instances would be associated with different partners. Skippers might well participate as owners of boats within larger groupings. Nevertheless, the ramification of the interests of the main owners—the first innovators and the salesmen seeking an outlet for their funds and a way of enlarging their businesses—did imply a substantial concentration of ownership in relatively few hands, although for particular vessels (as we have noted above) there would often be an association with minor partners. The right of acting as manager for a group of boats was often secured by part-ownership. Thus, a man such as Pyper would have the outright ownership of some vessels, a more widely distributed pattern of shares in vessels for which he acted as agent—that is both organizing sales and arranging for the

---

[1] *Lloyds List*, 1913.   [2] *Post Office, Aberdeen Directory, 1882–3.*
[3] *Post Office, Aberdeen Directory, 1892–3.*
[4] *S.C. on Sea Fisheries, 1893–4*, QQ. 6441, 6627.
[5] *Post Office, Aberdeen Directory.*

provision of stores—and, increasingly, an interest ramifying into other aspects of fishery supply.

After 1900, the fleet increased rapidly with an accompanying spread in ownership. More vessels came to be held in groups of one, two or three; and by 1913 well over half the fleet was so owned.[1] In the proliferation of these smaller units some of the trawl skippers rose either to full ownership of their vessels or to some form of direct participative share along with shore owners. By 1913 at least five skipper-owners can be discerned. But the most important movement of the first decade of the twentieth century was the advance of limited liability companies which operated groups of trawlers. The movement was particularly evident between 1900 and 1905, and by 1913 just under half of the fleet can be found in company ownership.

The companies drew much of their capital from the general public of the city of Aberdeen. The great majority of them made public issues of shares, and it is clear that they mobilized capital from people who had no practical interest in fishing. Thus, a good 90 per cent of the shareholders (at the first subscription) in companies formed between 1900 and 1911 were of occupations that had no evident connection with fishing.[2] In terms of the amount of capital subscribed the non-fishing interest was somewhat less predominant but was still much in the majority.

On the other hand the capital was drawn from a narrow area, the great bulk coming from the city of Aberdeen itself. The county of Aberdeen provided a much smaller amount but this was nevertheless still greater than the total from all other sources. In a very few cases this pattern of local predominance was shaken, as with Dodds Steam Fishing Company and the Red Star Steam Fishing Company which drew large proportions of their capital from England. But even in these exceptional cases local capital and local shareholders were still strong.

How far this wider subscription to trawling enterprises meant a shift in effective control is not so clear. Twenty-eight out of the seventy-two directors on the boards when the companies were first formed had no evident prior connection with fishing.[3] Perhaps more significantly, of the remainder who were in businesses which might conceivably benefit from fishing only two were described as fish salesmen, the trade which had traditionally dominated in trawl-owning; a much greater number were in fish-curing or other forms of processing. Some clue to the nature of the control within the companies is given by the composition

<hr>

[1] *Lloyd's List*, 1913.   [2] SRO, Board of Trade Register of Companies.
[3] Ibid.

of that evidently more powerful group of men who had seats on the boards of more than one company. These were thirteen in number and only eight were of trades closely connected with fishing, two being fish salesmen. A substantial proportion of the apparently powerful were of miscellaneous business experience.

The companies assisted in, and made more public, the knitting together of the various interests in supplies, ship-building and repairing, and processing that clustered around fishing. Thus, by 1913, of the sixty-five men who were either individual owners or directors of fishing companies, thirty-three can be traced as having direct interests in other aspects of the fishery business.[1] The commonest links were between owning and fish-selling and between owning and ice supply; there were eight individuals making one of these two types of combination. This was followed in importance by the link between curing and owning, with four instances (which, of course, meant that only a tiny proportion of curers had any direct interest in fishing itself). Other connections are to be found between owning and iron-founding, engineering, coal supply, insurance and ship-chandlery, all with three or fewer instances of the particular combination. The most common pattern was of the combination of trawl-owning with one other type of interest. An exceptional case of many interests was W. H. Dodds who was involved in ice manufacture and coal supply as well as owning independently one trawler and being the managing director of the Dodds Steam Fishing Company.

In 1919 the position was to be made yet more clear when directors had to register directorships in other companies.[2] It was then shown that every one of fourteen companies had at least one director with associated interests; furthermore, all but one were integrated by inter-locking directorships with at least two forms of associated enterprise. All companies were at least partially controlled by men whose profits might come not so much from trawling itself as from the activities associated with trawling. The commonest link was with ice supply, followed closely by coal supply. Links with insurance and curing and processing were also important, with ship-building and with general stores less so, at least as measured by the number of cases of linkage.

[1] *Post Office, Aberdeen Directory.*
[2] SRO, Board of Trade Register of Companies.

CHAPTER X

The Crofting Communities in the New Era, 1884–1914

I

THE CROFTERS' HOLDINGS ACT of 1886 had an inevitable, although uneven and partly unintended, influence on the fishing which was so close-knit a part of crofting life. In the fishing areas of Argyllshire, it is true, the fishing community had already largely split away from farming and crofting and in Shetland the tide of spontaneous development was running so strongly that legislation relating to land was largely irrelevant; but along the north-west mainland and in the islands where the fisheries were both closely interconnected with agriculture and delicately balanced between growth and decline, the Act was an appreciable influence.

Perhaps the most important result was its confirmation of smallholding as the basis of social life in the north-west Highlands. The Napier Commission whose report had provided the arguments for legislation had advised against granting security of tenure on the smallest holdings and had also looked forward to the separation of fishing from farming; but the legislation which followed granted security of tenure, fair rents and rights of bequest for even the smallest of rented holdings. The result was that the tenure of every tiny holder was reinforced, that the general run of holdings remained too small to provide independent livelihood and that most families continued to combine farming with some other means of earning a livelihood. Fishing tended to diminish somewhat, but survived largely as a part-time employment. However, the fact that the great bulk of the fishermen in the North West continued to hold land did not preclude a considerable development of fishing in the region. Even the most advanced forms of fishing, pursued with sailing vessels, were possible in the form of an activity occupying only a few months of the year; and the next thirty years were to see a minority of crofters greatly increase their catching power and fishing income. Till 1900 they seemed to be creeping up to achieve the efficiency of the east coast men who for so long had outclassed them in results.

But then came the steam drifter which could be worked only on a full-time basis; and the man who combined crofting with fishing, even if momentarily successful, began to fall far behind the full-time fisherman of the east coast in efficiency. Ultimately the crofter-fisherman was to be confined to fairly minor forms of fishing where inexpensive equipment could be used.

Some additional impetus was given to the effort to develop west coast fishing—still linked with working the land—by the use of public funds, to fill some of the gaps in the social and physical environment. One difficulty, which had become the more serious with the rise in the cost of boats and nets, was lack of capital. To meet this need, the Act of 1886 provided that the Fishery Board was to grant loans at low interest for the purchase of fishing boats.[1] By 1890, when lending stopped, some 246 loans had been made. Of these just over half—130— went to the west coast and particularly to Lewis and Barra. The average size of loans was £98—a smallish fraction of the cost of a new herring boat equipped for fishing, but a worthwhile contribution in acquiring one of the many east coast boats that were being sold second-hand at this time. The herring fleet of the Outer Hebrides was considerably increased in strength in these years and even the very modest provision of public funds must have been of some help. Then came the Western Highlands and Islands Act of 1891. This sought to provide help in the building of piers, harbours and boat-slips in the North West. In 1897 the administration was taken over by the Congested Districts Board. Only one full-scale harbour—at Ness—was built, but by 1910 the list of new and improved piers and boat-slips was a long one.[2] The list was convincing in its length but the full effects of numerous improvements can only be seen by an analysis of needs and difficulties inherent in the particular forms of fishing.

Furthermore, public help was only one fairly small influence in the many that were affecting the profitability of fishing. One of the major difficulties facing the fisherman of the North West was always the lack of a nearby market large enough to take an appreciable proportion of the catch in fresh condition. Nearly all landings had to be cured and the price of fish for this purpose was always well below the price of fish which was to go fresh directly to the consumer.

The problem was partly one of transport. After 1850 new methods of transport began to be developed which secured the speed of transit necessary for reaching the main markets with fresh fish from the North

[1] FB Rep., *1899*, 171.
[2] Congested Districts Board, *Report, 1907–8*, 16.

West. The first step was to establish a regular steamship connection. By 1865 Stornoway was sending about a quarter of its herring catch southwards uncured, but this seems to have been almost the only point from which regular consignments were made.[1] Then, in 1870, the railway reached the west coast at Strome Ferry.[2] This railhead was close to one of the main fishing areas of the North West, but the configuration of the loch on which it stood resulted in problems of shipment and access from the open sea and it gave at best only a roundabout connection with the main railway network of the country, by a line which reached the coast, unavoidably, at only one point among the many and widely spread lochs and fishing bases. The problem of developing a system of quick land communication with the general market areas, connecting with the multitude of places where small boats might land their catches, was considered by two main government committees, in 1890 and 1905. A network of railways, which would necessarily have to radiate from focal points in the east, was discussed with approval.[3] In fact the sole outcome was the lengthening of the existing line from Strome Ferry to the better landing point of Kyle of Lochalsh and, much farther south, the completion of the line between the coastal terminus of Mallaig and Glasgow.[4]

Many problems of transportation remained to be solved. Freight rates were heavy and the time involved in transit was still considerable, while most of the fishing areas were still far removed from the Highland railheads; but by use of additional transport devices, a considerable proportion of the catch was funnelled to the two railheads to be sent southwards as fresh fish. The developing use of the steamship in conjunction with the railway provided a means by which fish, caught many miles from the railhead, could be carried by steam from the extreme corners to the markets of southern and midland England. The railways then had an appreciable effect on prices, even of the (still considerable) part of the catch that had to be cured within the area. Again it is only by investigating the varying effects on different areas and for different products that the impact is to be measured.

Against this background, some of the problems that had kept the loch fishing for so long uncertain and trivial began to ease. Boats of greater efficiency and size began to be acquired in numbers. They were

---

[1] *Sea Fisheries Commission, 1866*, Q. 32618.
[2] The Stephenson Locomotive Society, *The Highland Railway Company, 1855-1955* (London, 1955), 26, 33.
[3] *Committee on Fisheries in Sutherland and Caithness, 1905*, Q. 2919.
[4] Stephenson Locomotive Society, *Highland Railway Company*, 26, 33.

mainly of Loch Fyne type, half-decked and built to a length of twenty-five feet, and they could more easily be moved over relatively long distances to those shifting areas where there was plentiful fishing.[1] Morover, they would carry to these fishings a much more complete and efficient set of nets than had commonly been found among the west coast boats in the past. While the local catches and even the regional aggregates for the annual fishing both continued to fluctuate greatly, the individual result became more certain.[2] Any reasonably equipped crew could take some share in the still fluctuating catch of the whole region where, in the past, it had been common for some crews to suffer completely blank years. Still, greater mobility and correspondingly keener response to fishing opportunity were achieved by some crews by the adoption of auxiliary paraffin motors; by 1910 a significant minority of the boats had been fitted with motors.[3] The results were to be seen in the much bigger yields of the improved units. Thus in 1911 the motor-boats averaged gross earnings of £200 as against £70 for the sailboats, and in 1913 the corresponding averages were £300 and £90.[4]

The use of better equipment was partly the result of the opportunity provided by the cheap and portable auxiliary engine but also it was the result of considerable investment in instruments of fishing that had long been available. Investment depended not only on changes of attitude but also on an environment that would allow a reasonably calculable gain to each increment of expenditure. Higher and steadier prices, coupled with greater certainty of catch stemming from the increased mobility, raised incomes in a way that was directly traceable to the use of improved equipment. Price was steadied by the better co-ordination of curing with fishing and by the strengthening, by means of the railway, of the links with the consumer. Once the fishing in a particular area, was shown to be promising a miscellaneous fleet would converge to bring in curing stock and to carry away landings uncured. Sloops with curing stock would arrive; curing stations would be set up on shore; and some merchants would complete the curing process on board ship (although this traditional form of curing was tending to diminish). The new element in the system was the use of steamers in considerable numbers. Sometimes twenty or thirty would collect in the one loch, to carry the fish either directly to the main markets in the south or to the railhead from which they would go, lightly sprinkled in salt, to the industrial

---

[1] FB Reps., *1897*, 222; *1901*, 246; *1910*, 226; *1913*, 226; *1914*, 232.
[2] FB Reps., *1909*, 230; *1910*, 227; *1913*, 228; *1914*, 233.
[3] FB Reps., *1910*, 226; *1913*, 226.
[4] FB Reps., *1911*, 227; *1912*, 222; *1913*, 228; *1914*, 233.

districts of Scotland or even to England.[1] Both steam carrier-vessels and larger fishing boats contributed to the greater centralization of curing, since wherever the fish were initially caught they might be carried to a centrally placed curing yard; or alternatively the fishing vessel itself might run to a base at some distance from the fishing ground. Thus fishermen were made more sure of a market for their fish; and curers, less tied to every tiny local circumstance, were more sure of a supply to keep their yards busy. On the whole, then, it became less likely that landings would run to waste because of lack of buyers. Since there were several ways of dealing with the catch, the system became more flexible, and total capacity grew.

With steadier, and on average higher, prices for fish, with the increase in the ability of crews to move into the promising areas of fishing, and with a growing capacity to catch once they had got there, the possibility of disastrous failure of the loch fishing diminished in general and for the particular crew. Most fishermen now shared in the fishings of the region as a whole, so that, while incomes continued to vary from year to year, both the upper and lower limits of this fluctuation rose significantly in the last decade of the century.

Throughout the nineties, boats engaged in the loch fishing would make yearly averages—in gross takings—of between £30 and £50. This meant individual earnings of less than £10, against which there had to be set considerably increased capital charges. Even greater than the variation from year to year was the difference in the success of the different boats within the one year; in 1898, for example, boats in the Loch Carron district made from £20 to £80 each. After 1900 earnings rose appreciably, due both to the addition of the motor boats which had so much greater earning power and to an upward trend in the catches of the boats which had not been adapted. Thus the average earnings of motor boats stood at £200 in 1911, while in 1912 and 1913 they made between £300 and £500 and in 1914 the top earnings reached £640.[2] Over the same years, earnings of sailing boats rose from the average of £50 in 1911 to £90 in 1913 and £100 in 1914. The years after 1910 may have been a chance cluster of a few successive years of good fortune, but the trend during these years was upward. Certainly no previous similar group of years had shown the same success.

Even earnings around these levels do not mean that the full potential of the easy fishing conditions of the lochs was being realized. The improvements in the marketing mechanism did not fully solve the

---

[1] FB Rep., *1900*, 265.
[2] FB Reps., *1911*, 227; *1912*; 222; *1913*, 228; *1914*, 233.

problem of prices and there were areas where the costs and uncertainties of reaching one of the points at which curing yards were busily at work held prices considerably below the general regional level. It is notable that the existing level of prices stood lower the greater the distance from the railheads or from the main established centres of curing.[1] Thus the west of Skye had lower prices than the south-east corner of the island with its proximity to the railhead at Kyle.

Even at the railhead prices were not as high as they were at points nearer to the points of consumption. The best-placed fishing area of Skye still did not have anything like the market advantage of, for example, Loch Fyne (with Glasgow nearby). There remained a considerable margin between prices in the main markets and in the North West because the costs of transport were so high, this burden being particularly heavy on such cheap fish as herring. The lower rate at which herring could be carried from Strome Ferry to London was 3s. per cwt., a rate applying to lots of at least three tons, carried by slower train. Prime fish despatched by passenger train in small lots was rated at 4s. 9d. per cwt.[2] Even the lower rates represented an addition of two-thirds to what was paid by the merchant to the fisherman.

In the late eighties and in the nineties the price of herring on the north-west mainland was still well below that in the Outer Isles, as paid by east coast curers for herring to be delivered to their yards during the early and main summer seasons. Thus on the average of the years 1889–92, prices in the Loch Broom and Loch Carron districts (which between them cover the whole mainland north of, and including, Skye) were slightly over half those prevailing in the Outer Isles and not much more than a third of the Campbeltown values. By 1910, however, there had been a levelling within the different areas, and in the case of the north-west mainland considerable increase. In Loch Broom the price over the period 1889–92 had been 9s. 7d. per cran while on the average of the years 1910–13 it was about 16s. per cran, exactly the same as the average for the Outer Isles and not very much below the Campbeltown level.[3]

A second persisting difficulty in the loch fishing was the continued fluctuation of the general average of catch for the whole region. Fishermen might escape from the misfortunes of a particular locality by their new mobility but they had to accept the impact of the variations that struck the region as a whole. The figures for landings in the fishery districts do not fully account for all the catches of the boats of the

---

[1] FB Reps., *1893*, 182–3; *1894*, 182; *1903*, 231.
[2] *Fish Trades Gazette*, 11 Feb. 1888.   [3] FB Reps.

region but they are probably significant of variations in the level of landings.

It is unlikely that these fluctuations were greater than they had been earlier in the century but their consequences seemed to be growing more serious. A few years of failure or of low catches would be followed by the permanent laying-up of boats. Partly, this was due to the fact that men who were only casually engaged were discouraged and gave up more readily than once they had done. When fishing was carried on for subsistence purposes, the landing even of enough fish to fill one barrel was a significant contribution to the family's comfort; and the fishing would continue when there was any chance of this minimal reward. But, with the growing reliance on purchased foodstuffs, men who had found it worthwhile to keep some equipment in order to obtain trivial supplies from the loch would now do so only in the hope of a substantial money income. Investment in the upkeep and equipment of boats had become much more sensitive to the return from fishing because the reward had to be more definite. In addition, the wages that could be earned in various types of seasonal work were now so much above the return to be expected from the occasional fishing venture that they tended to look for such work rather than to engage in the fishing.[1]

Without increasingly expensive apparatus a crew's earnings remained desultory and trivial; and only a diminishing minority were ready to expend the continuous effort, and to meet the expenses of preparation, necessary to make such earnings. A Loch Fyne skiff cost over £100 as compared with the £10 that had been the traditional investment in a boat for the west coast loch fishing and an auxiliary engine would add another £110 to £130, so that when nets were included it would take over £200 to buy and equip fully a boat for the loch fishing.[2] In west coast communities, then, some men, finding that the trivial and uncertain rewards of a desultory and badly prepared fishing were not worthwhile, tended to give up fishing altogether; others—a minority—made the investment (and the concentrated effort, often involving long journeys) that brought considerably rising incomes. The earnings' figures, in fact, pertain to a diminishing minority; but even this minority held on to their land and were not to be found in concentration in any particular area.

Fluctuations in fishing returns further diminished the numbers of those able to keep up with the pace of investment. The blight of the poor years, falling on specific districts, would cut out those crews with

[1] FB Reps., *1903*, 238; *1905*, 218; *1907*, 283; *1910*, 226; *1913*, 226.
[2] FB Reps., *1910*, 226; *1913*, 226.

inadequate equipment and would make it difficult to effect the improvement in equipment needed for great success. The result was that, as boats became larger and more expensive, they also become fewer in numbers. Thus between 1890 and 1913 the number of boats in the Loch Broom district declined from 602 to 351 and in the Loch Carron district from 962 to 397. Of these it was the largest and the smallest that showed the greatest fall in numbers—the small boats because they were now fairly useless in serving the social purposes of the crofter and the large because they had been mainly used in the Minch fishings from which the mainland fishermen were withdrawing. In so far as the largest class of boats declined so steeply, the general fall in numbers of boats probably understates the decline in the numbers engaged in the herring fishing. Fishing, in fact, was narrowing down to an operation by boats of the Loch Fyne type, of 20- to 25-foot keel. In the Loch Broom district in 1879 there were only sixty-eight boats in the general category that would include this class and ninety-four in 1913, while Loch Carron had respectively 329 and 171 at the two dates.[1] Such figures would indicate that a relatively small proportion of the population was making any true effort at the fishing, although it could do so with improved effect.

II

The line fisherman of the North West showed less power to adapt than did the herring fisherman. Line-fishing, mainly for cod and ling, had a well-defined localization within certain communities in which it took a much more definite place in the fishing calendar than did herring fishing; but development was difficult and, in fact, in a period when the other forms of fishing were changing with some rapidity it remained largely stagnant in scale and method. Indeed, just before 1914, all activity of this type died away because it was proving so unrewarding in comparison with other forms of work.

Throughout the nineteenth century this was a type of fishing of an unused potential, as was shown by the incoming crews from the east coast when they were able to make as much in a single trip to the fishing grounds as could the local men within a whole season, bound as they were to the limits of their traditional fishery. The main basis of progress had to be the adoption of bigger boats which would give longer periods at sea, more regular fishing, and the landing of much larger hauls obtained by using better outfits of lines. One obstacle to the use of bigger boats was lack of capital. Even for their 20-foot boats many

[1] FB Reps., *1878*, 23; *1890*, App. D, No. VIII; *1900*, App. A, No. II; *1913*, App. A, No. II.

line fishermen were deeply in debt to the merchants, and some were not even nominal owners.[1] But this indebtedness was itself partly the result of the vicious circle of poor equipment stemming from low yields which gave no surplus for investment in better equipment—a circle that could be broken by a bold decision to invest in larger boats, self-justifying in producing yields out of which loans could be repaid and fresh savings made.

In fact the restraint on improvement imposed by lack of capital was rather unimportant because of another difficulty which set a much more definite limit to any increase in the size of boats. The lack of sheltered water at many of the points from which the white fishers worked their boats meant that no craft could be used which was too heavy to be hauled over the beach.[2] To create harbours at all of these points would have been an impossibly expensive project, unlikely to be met by any normal mode of harbour finance. The beaches were in many cases entirely open and protection would have to be accorded by the harbour structure itself. The programme of building piers and boat-slips which made some progress after the Act of 1891 was adapted to waters which had some degree of initial shelter. But to create entirely artificial structures at a large number of points, in each case for the service of only a small number of intermittently used craft, implied lavish expenditure; and only a massive grant of public funds, with little prospect of repayment out of newly generated traffic, would have met the case. In fact, one artificial harbour was built at Ness on the extreme northern point of Lewis in the hope that in a region with a scattered but numerous fishing population some boats would be drawn from a distance to work from that one good harbour.[3] In the event, the Ness plan was a failure for two reasons. Firstly, a series of engineering difficulties held up construction and also made the finished port liable to silting—and therefore insufficiently convenient to draw boats from a distance. Secondly, white fishermen showed themselves unwilling to make even short moves from their home base. This was due more to the particular requirements of the fishing, which was intermittent through a long period and which made use of family labour, than to the obstruction of personal and

[1] *Commission on Crofters and Cottars, 1884*, QQ. 17519, 15897-8.

[2] FBR Report on Shawbost, 30 Nov. 1880; Petition from Fishermen of Shawbost, 1880; Report on Ness and other harbours on the West Coast of Scotland by Dugald Graham, 1883; Report on Lewis Harbours, 10 Oct. 1885; Report by the Fishery Officer (on harbours), 10 Oct. 1885; *Commission on Crofters and Cottars, 1884*, QQ. 14391, 15454, 15816, 16110, 18071-2, 28026, 28913, 29294, 29318; *Fish Trades Gazette*, 10 Apr. 1890; FB Rep., *1892*, p. xxxix.

[3] J. P. Day, *Public Administration in the Highlands and Islands of Scotland* (London, 1918), 267-9.

fundamental attitudes among the fishermen; for many of them, after all, were prepared to make long migrations, with or without their boats, in pursuit of employment in the herring fishing.[1] In fact a whole way of life was built around the line-fishing in which the concentrated labour of the whole family was articulated to make possible a fishing at difficult positions. A technically similar type of fishing (but without the family ancillary) implied the breaking-up of complex traditional procedures in which fishing was related to the whole life of the township, and to remove the fishing from the old context meant readjustment in more than the simple operation of fishing itself.

The cod and ling fishing in its traditional form proved incapable of gradual adaptation and it had to survive in its general traditional form or die. But within the traditional context it was impossible to introduce the detailed improvements that would make it a tolerable manner of earning a livelihood in the circumstances of the twentieth century. So it died.

Another problem besetting line-fishing in the west coast areas was that of prices, which were generally lower than elsewhere because, owing to poor transport, the great fish had to be used for curing rather than go for sale as fresh fish. Thus in 1889–92 the average price of cod in the Loch Broom district was 4s. 9d. as compared with 8s. on the east coast and 7s. 9d. at Campbeltown; the Outer Isles price of ling—the catch of most importance to the area—was 5s. 7d., as compared with 8s. 7d. paid on the east coast.[2] And this was at a time when the railway and steamship had already given some lift to west coast prices. The Loch Carron values, for example, were above those prevailing farther from the railhead. Marketing was likely to be eased more in the case of cod than of herring, since cod had a higher value in relation to its bulk and was also, in unprocessed form, less liable to deterioration. Thus the price of cod at the railhead was raised but the benefit failed to spread through the whole region of the North West and in particular it by-passed some of the important points of line-fishing. The steamship, in this case, was a rather ineffective adjunct to the railway. With cod landings occurring intermittently through the week, small quantities had to be uplifted at frequent intervals from a large number of points. Heavy landings of herring, when they did occur, were generally concentrated at sites which allowed regular services to be organized, whereas the cod and ling were landed at many different points and so defeated any attempt to organize a regular traffic. Few of the points at which cod or ling were landed had, individually, sufficient supplies to justify the

[1] FB Rep., *1899*, 73–5.   [2] FB Reps.

running of a service to the railhead frequent enough to provide a connection on all days when landings were made.

Until 1905 the best that could be done north of the Kyle railhead was to provide a steamer which called at the various places on its long round, once a month in winter and once a fortnight in summer, and even with this schedule the calls were somewhat undependable as well as infrequent. Some points still found themselves without nearby services and, for the calls on the outward half of the run, a long time had to elapse before the consignments would reach the railhead. In 1905 the Congested Districts Board tried to start more regular and direct services by offering a guarantee against loss, for stipulated trial periods, to a steamer which would run on two routes—one between Kyle and Ullapool and the other farther north to Sutherland.[1] The traffic generated on the second run was far below that required for commercial operation and the service was removed after one season, but the link with Ullapool proved worthwhile and was continued on an independent basis. It was part of the reason for this success that it provided a market outlet for the cod landed regularly at Badachro, near Gairloch, some thirty miles to the north of Kyle. About thirty boats came to be regularly employed at Badachro, making seasonal gains of up to £100 each, which was somewhat greater than the income from line-fishing anywhere else in the North West.[2] This was, indeed, the only place in the region where a local cod or ling fishing was successfully linked with the railway. Fish landed in the other specific areas of line-fishing—in Lewis, Barra, Sutherland and Skye—continued to be used mainly for curing. Prices rose in all areas between 1890 and 1910 but the outlying parts of the North West, where so many of the line-fishing communities were situated, continued to be at some disadvantage relative to the improved conditions at the railhead.

In the outcome, apart from the crews who were making good fishings and good prices near the railheads and apart from some who made for the market centres of Caithness, cod and ling fishing persisted in a purely traditional form, making no gains from the changes of the later nineteenth century. With gear and methods that had scarcely changed in essentials since the end of the eighteenth century, there was little scope for raising incomes. Yields were low and, while prices drew nearer to the levels of the east coast, the gap was never fully closed; and the conditions of fishing remained as hard and dangerous as ever. The

[1] Congested Districts Board, *Reports, 1904–1905*, p. xx; *1905–1906*, pp. xxxiv, xxvii; *1906–1907*, pp. xv–xvii; *1907–1908*, pp. xvi–xvii.
[2] FB Rep., *1910*, 227.

consequence was a revulsion from an unrewarding pursuit and the gradual decline of activity, till by 1914 there was almost nothing left of the traditional line-fishing. The old open boats had gone, the beaches were deserted and the fishermen were engaged either in quite different pursuits or in fishings of other types.

In the case of Sutherland, there seems to have been decline in the yield from line-fishing, generally blamed upon the operation of trawlers;[1] but elsewhere there is the puzzling picture of a fishing maintaining its catch, selling at slightly increased price and yet going into irreversible decline. The explanation seems to be that, as an occupation in traditional form, it gave incomes which rose too slowly to hold its labour force when standards of living and expectations about standards of living began to rise. It was possible for great-line fishermen, if they were willing to operate in the different environment of large fishing centres, to make incomes which would compare with other opportunities of the time. But few did so. They found it easier to adjust by entering on entirely new types of fishing or by taking up entirely different forms of employment. In Barra and Lewis it was probably the incomes to be made in herring fishing that led to the decline of the local white fishing.[2] In Skye, we find fishermen taking jobs as crewmen in yachts.[3] In Sutherland, the yields of white fishing collapsed even while it was becoming more difficult to maintain a position in herring fishing because of the increased expensiveness of equipment. As the men took to engaging as wage-earners on east coast boats, the maintenance of the boats solely for a failing white fishing became impossible.[4] The fishermen of an area which had depended more completely upon fishing than any other in the North West found themselves solely dependent on what they could make on the boats of others. In less than ten years the line-fishing of the area was swept away.

III

The Minch fishing—in which local boats were embodied in the fleet of east coast boats to work on behalf of east coast curers from the Long Island station—was little affected by the transport improvement of the late nineteenth century. Curing stock continued to be carried by sea, as it had been since the 1840s, to stations that were fixed for the season;

[1] *Committee on Fisheries in Sutherland and Caithness, 1905,* QQ. 2215, 2229, 2597, 2916, 3043, 3107, 3248, 3422.
[2] FB Rep., *1903,* 87; *Fish Trades Gazette,* 10 May 1890.
[3] FB Reps., *1905,* 218; *1906,* 269; *1910,* 208; *1913,* 226.
[4] *Committee on Fisheries in Sutherland and Caithness, 1905,* QQ. 2597, 2666, 2706, 3138.

all necessary labour was accommodated on the temporary stations; and the finished cure was exported directly by sea either to the Continent or to Ireland. This was the pattern that continued to prevail till 1914, with the slight improvement, affecting east coast as well as west, that the curing stock and the finished cure were carried in steam rather than sailing vessels. But the coming of steam did not in this case mean the opening of a new channel of communication as it did for much of the other fish caught in west coast waters. Rather more of the herring came to be sent south as fresh fish, kippering developed as a major enterprise taking some of the landings at Stornoway, and vessels began to arrive from the Continent to carry away the herring uncured; but the great bulk of the catch continued to be cured by pickling. The system by which boats were engaged to work for particular curers was somewhat more durable on the west coast, at least in the remote parts, than it was on the east; but by 1887 Stornoway was operating on an auction system and between 1900 and 1905 the system of engagements gradually gave way in Barra.[1] However the fish might be sold to the curer, the basic determinant of price was the condition of the Continental market and the price of cured fish that ruled there. Other changes were of comparatively minor importance.

West coast crews, then, were just as affected as any by the great crisis of 1884. The result was an immediate severe drop in price, reflected in engagements running at 15s. per cran in place of the 20s. that had been standard before the crisis year.[2] In addition, fewer boats were prepared for the fishing and the west coast men suffered because they found it more difficult to gain employment as hired hands since the east coast boats were increasingly being crewed by fishermen of their home region. Both because of the drop in wage-payments and because of the failure of local boats to make so much of the fishing, the depression of the late eighties severely hit much of the west coast.

The recovery which had begun to show itself by 1894 was moderate and never carried the industry to the heights of profitability that had been known before 1884. Prices for both raw and cured fish failed to reach the highest levels of the seventies and eighties. Nevertheless, at this restrained level of profit the herring industry as a whole was to continue to expand without serious recession till 1914. The effects of this expansion intermingled with the effects of serious social changes that were being forced on the industry. Between 1890 and 1900, when the building of boats began to recover from the depression, there was a

---

[1] FB Rep., *1911*, pp. xlii–xliii.   [2] *Fish Trades Gazette*, 30 Apr. 1887.

steep increase in the price of the standard unit and by 1900 it cost £1,000 to build and equip a herring boat.[1] Thus by the turn of the century an increasing number of men who previously had been owners of boats were being forced into the position of wage-earners; sometimes they would contribute nets alone but many lost any property stake in the fishing.[2] Then after 1900 the building of new sail boats for herring fishing virtually ceased as all available capital went into the steam drifters, and it became harder than ever for those without a substantial existing share in the industry to acquire ownership. These were general forces but they applied with particular strength to the west coast.

The acquisition even of sailing boats of the bigger type that was becoming common and necessary in the last decade of the century proved to be beyond many west coast fishermen who had previously held shares in herring boats. Burdened with debt and accustomed to buy second-hand boats of less than full efficiency, they had little scope for saving and little credit standing, and they had always found the next upward step in equipment difficult to make. Thus the soaring costs of equipment and ownership in the nineties put them out of the running and reduced them to the status of hired men.[3] Another result of the use of larger boats was that many communities without harbours were now, even out of season, forced to maintain their boats at a relatively distant deep-water port, whereas previously they had simply drawn up their boats on the beach at home ready for the season's moves. It was not perhaps a decisive disadvantage—some of the east coast villages managed on such a basis—but when added to other encumbrances it might deprive the fisherman (who had formerly been a struggling owner) of a full share in the fishing. These forces, squeezing out the less affluent and successful from the position of proprietors, became much stronger with the advent of the much more expensive steam drifter. In particular the west coast men were virtual strangers to the merchant community— since the curers ceased to be the main backers—from which the financial aid must come. Thus, while the steam drifters were swarming around the west coast, as late as 1914 only one was owned by a west coast crew.[4] Yet the rush of the east coast men to invest in steam drifters did bring one advantage to the west coast men, since second-hand sailing vessels of recent design came on the market at low price. It became easier after 1900 to sustain a position with a sailing vessel than it had been in

[1] FB Rep., *1900*, App. G, No. 1.
[2] *Committee on Fisheries in Sutherland and Caithness, 1905*, QQ. 703, 822, 1422, 3383, 3587, 3681.
[3] See above, pp. 97–8.
[4] *Committee on North Sea Fishing Industry, 1914*, Vol. II, Q. 2618.

the nineties; and the fishermen who had not completely lost their position as owners were able to continue in possession of sailing boats which, by any standards other than those of the competing steamship, were of unprecedented efficiency. Further, up to 1914, the market for cured herring was expanding fast enough to allow the sailing boats to fish alongside the drifters without glutting the market. The industry, therefore, still had a place for the sailing vessel and the crews who were in a position to buy boats between 1900 and 1914 could use their vessels profitably both on west and on east coasts.

In fact, the participation of the people of the north-west mainland had been irretrievably destroyed before the beginning of the steam drifter era. In 1870 the crews of north-west Sutherland had been heavily committed to herring fishing and had been making some success of it: but then there began a decline which steepened in the eighties and nineties.[1] By 1900 the participation of the north-west mainland in the Minch herring fishing was merely nominal. Whereas the Loch Broom district had 125 boats of over 30-foot keel in 1855 and 74 in 1879, it had but 27 in 1900.[2] The later boats were much bigger and more effective, but they required at most one additional member to the crew, and the fall in the number of boats accurately indicates the similar fall in the number of fishermen participating—as owners of boats rather than as hired men—in the herring fishings of the Long Island and the east coast.

The experience of the Outer Isles, and particularly of the eastern coastline of Lewis and Barra, was different. In the nineties, the number of vessels of more than 30-foot keel owned in the Stornoway district did diminish but, with 88 out of 158 being of more than 45-foot keel, the district was evidently investing in a new generation of boats.[3] In the Barra district the fleet of larger boats grew in size, although the proportion of crews with boats of the largest size was less than in Lewis. After 1900 the Stornoway fleet of large boats diminished somewhat and then, by 1914, recovered to number 122.[4] In Barra there was a more persistent increase; it was here that most use was made of the supply of second-hand boats released from the east coast by the re-equipment with steam drifters. The fleet increased from forty-one to sixty-six between 1900 and 1913 and, as in Stornoway, the individual units came to be used more intensively, with a tendency to carry on the fishing

---

[1] *Committee on the Employment of Women in Agriculture, Second Report, 1870*, 324–7.

[2] FB Reps., *1856*, 24–39; *1879*, 23; *1900*, App. A, No. II.

[3] FB Reps., *1890*, App, D, VIII; *1900*, App. A., No. II.

[4] FB Reps., *1900*, App, D, VIII; *1914*, App. A., No. II.

through the summer when the fleet of incomers from the east coast had departed.[1]

The individual incomes of those who managed to hold on to their boats, even if they were only sailing boats, rose markedly between 1900 and 1934 and were generally much above incomes earned in the traditional line-fishing. A boat engaged in the line-fishing only exceptionally grossed £100 a year and usually the earnings would be between £40 and £70; but in herring fishing, while catches fluctuated a good deal from year to year, the annual average for the main early summer fishing never fell below £100 and sometimes approached £200. There was also each year a big variation as between the different boats and, while the gross earning of some boats might be as low as £50, others would make £300 or more.[2] When fishing was continued through the summer on the west coast a somewhat smaller sum would be added to the boat's earnings. But the boats that went to the east coast for the main summer season would probably more than double their initial takings on the west coast. It was the Lewis men, rather than those from Barra, who normally made this long migration and one or two crews even extended their season further by visiting East Anglia in the autumn. The most successful Lewis crews, then, were beginning to rival the east coast men who still worked sailing boats, with annual gross takings close to the £1,000 mark (a return which was in fact reached by one crew as early as 1900). But the journey to East Anglia was made only by a few crews and the usual ceiling of a boat's gross takings would be £500. The herring fisherman had much greater expenses than the line fisherman and he had to make greater allowance for the depreciation of his expensive equipment, but certainly the incomes that were being made by the men from Barra, and even more by those of Lewis, were far and away greater between 1900 and 1914 than any they had previously known from fishing.

These incomes were made by a group that was rising in numbers. But, even so, this fishing with big boats which had to be used for several months of the year was a minority occupation in the islands. The decline of line-fishing, involving the life of the whole township, had tended to reduce the fishing base of island life. While a minority were making good incomes, the majority who had once combined crofting with fishing turned now to other pursuits. All over the North West, then, fishing was shrinking from being the general employment of a great mass of the population to surviving as a small specialist occu-

[1] FB Reps., *1900*, App. D, VIII; *1913*, App. B, No. II.
[2] FB Reps., *1900–1913*.

pation in which participation was derived from a fairly considerable investment—an investment which was only worthwhile if it was followed by an annual and concentrated pursuit of fishing lasting over several months. This was true positively of both loch and Minch fishings, where bigger incomes were made with bigger boats but by a diminishing minority; it was true negatively both of line-fishing where decline meant the obliteration of a fishing which had fully engaged whole communities and of those areas where large-scale herring fishing had ceased entirely. Only lobster fishing was growing, for in it conditions of simple and cheap equipment combined with intermittent activity to allow fishing to remain as the true accompaniment, on a general scale, of crofting.

IV

As the nineteenth century wore on, the Clyde area (including the long eastern shore of the Kintyre peninsula) became ever more clearly different from the rest of the west coast in the regular industry and success of its fishers. With good prices and well-sustained catches a substantial number were able to live largely upon the proceeds of fishing. By the 1880s the upper part of Loch Fyne still held a few scattered groups of crofter-fishermen but farther south it was the full-time fisherman, living in a considerable community with others of his kind, who had become typical. Indeed, the boats and the fishermen were now mainly concentrated in the string of villages along the eastern shore of the Kintyre peninsula—in Ardrishaig, Lochgilphead, Tarbert, Carradale and in the largest concentration of all at Campbeltown. In Tarbert there was a memory of a time when crofter-fishermen had been common but it was so vestigial as to be only an unclear recollection.[1] The basis of livelihood was the herring fishing of Loch Fyne and the waters farther south, particularly the Sound of Kilbrennan between Arran and Kintyre. This fishing carried on in the whole complex of sounds and lochs, started in May and could last till the end of the year so that the participating crews would move from one section to another as seemed to be advantageous. For the first three months of the year also, some crews would go to Ballantrae for the herring fishing of that area, and sometimes there would be small migrations into the North West and particularly to Lochboisdale where there was a fishing in April.[2] Herring might be caught either by ring-nets which had been legitimized in 1867 or by drift-nets, and many crews still had the equipment for either

---

[1] *Commission on Crofters and Cottars, 1884*, QQ. 46251.
[2] FB Rep., *1894*, 189.

method. While some had ring-nets only and a few, possibly, had only drift-nets, the tendency was for the ring-net to take over. Generally the boats working ring-nets came to outnumber those working drift-nets by about two to one and sometimes the concentration on ring-netting would be even more pronounced.[1] The growing reliance on ring-nets coincided with another shift since, after 1905, the upper part of Loch Fyne, always an area exploited by drift-nets, yielded less and less to the fishermen.[2]

The fishermen of the region were all independent, both in their manner of operation and in the ownership of their boats and gear. Catches were mainly sold, by a process of daily bargaining, to the merchants on the 'screw' vessels which arrived early every morning to carry the boxes rapidly to Glasgow. A small and diminishing residue of the catch was sold to the curer but there was never any form of engagement. When crews went to Lochboisdale, however, they operated under the terms of engagement normal to that region. Boats and nets were owned entirely by the fishermen themselves, the proceeds being divided with fixed shares assigned to the boat and the nets, and the remainder was split among the whole crew as individual members.[3] There was much buying and selling of boats or shares in boats, but neither this nor the original financing of construction resulted in ownership moving into the hands of landsmen; indeed there was not even much debt weighing on the fishing community. The boats were the result of a slow development of the Loch Fyne skiff, which had always been well-adapted to its tasks and by the 1880s the standard size was of 26-foot keel, with the cost of a new boat just over £100.[4] This relatively cheap equipment was notably efficient for its task of ring-net fishing in enclosed waters. The continuing ability of the full-time Firth of Clyde fisherman to provide his equipment without tying himself to outside agencies, and at the same time to maintain a reasonable level of earning, was the outcome of the fortunate conjunction of good markets and cheap equipment which could be operated at a high level of efficiency.[5]

The tendency between the mid-eighties and 1914 was for drift-netting to decline even from the secondary position it had occupied. There were some switches from year to year and, even in mid-season, from one method to the other, but the drift-net was generally being abandoned. Almost complete specialization on ring-net fishing accompanied, if it was not caused by, a tendency for the fishing to move

[1] FB Rep., *1889*.   [2] FB Rep., *1906*, 274.   [3] FB Rep., *1911*, p. xix.
[4] *Commission on Crofters and Cottars, 1884*, QQ. 46287, 46321.
[5] FB Reps, *1902*, 248; *1904*, 235.

out of Loch Fyne itself into other parts of the Firth of Clyde, and indeed
beyond the region entirely. It was pressure of declining yields rather
than the attraction of great fishings elsewhere that caused the movement. In 1905 when there started a swift and protracted decline in the
herring fisheries of Loch Fyne, even the fishermen living on the loch
itself had to look elsewhere for their livelihood.[1]

The decline of the fishings in Loch Fyne itself was disastrous for the
remnants of the crofting-fishing population at the head of the loch and
the old mode of earning a living finally disappeared. But the full-time
fishermen living in the lower part of the loch—the men who had been
so prosperous in the 1880s—had greater resilience and found compensation in two ways. Firstly, they were able to move, in the course of
their seasonal fishing, to the areas where the fishings were better. Such
movement was no new thing; the change was rather in the balance of
time spent in home and in distant waters. The second compensation
was the fitting of paraffin motors to the skiffs. The first stage was to use
one motorboat along with one sailing boat in the pair needed to manage
a ring-net. This in itself nearly doubled the yield but the linking of two
motorboats brought another significant increase of catch.[2] Thus fairly
small expenditure brought a substantial improvement in earning power
and, even though there was a fuel bill now to be met, an engine might
be expected to pay for itself within the course of a single season. Such
clear advantages brought a swift changeover to motor power starting
in 1908.[3]

While the Loch Fyne men could partly salvage their fortunes by
adaptation, those of Campbeltown could obtain the same advantage
from motor power without being faced by the problem of a decline in
their local fishings. So quickly did they adapt their boats that by 1911
there were no pairs of sailing boats left.[4]

Time had, thus, brought the concentration of the fishing population
in a few communities where the possession of land was but a memory,
if that. At the same time total numbers of fishermen in the region were
declining; indeed, in due course the tendency for numbers to decline
spread even to the points where the fishermen had mainly collected. In
1899 Tarbert, Ardrishaig and Lochgilphead had, in all, 596 fishermen;
but by 1913 they had only 320.[5] The number of boats in the fleet, too,
tended to fall. The three places had 175 boats of over 18-foot keel in
1899, a number which had fallen to 102 by 1914. The boats themselves,

[1] FB Rep., *1906*, 274.   [2] FB Rep., *1911*, p. xv.
[3] FB Rep., *1911*, pp. xv, xix.   [4] FB Rep., *1911*, p. xv.
[5] FB Rep., *1913*, App. A, No. II.

however, were capable of ever greater catches. By 1914, virtually all were motorized, and by that time a new and superior class was being brought into operation. A boat of this latest generation was of 29-foot keel and 40-foot overall and would cost £275 for the boat and engine, the ring-net adding another £40.[1] This rise in capital costs was making it more difficult for all members of the crew to acquire shares, a tendency that was perhaps strengthened by a bias within the system of sharing which concentrated rewards in the hands of men already owning boat or nets; in so far as the hired hand took a minor share, it was correspondingly difficult for him to acquire greater standing. For the hired hand £2 per week was regarded as moderate earnings and the owner, even when he had paid all necessary expenses, might take twice as much.[2] The modest Loch Fyne skiff seemed to be providing as much return as the expensive steam drifters of the east.

V

In the early eighties the place of fishing in the life of Shetland was abruptly and deeply changed, as the traditional haaf system, which had occupied so many of the people for so long, suddenly gave way. It was replaced by a fishing, more narrowly based but in much larger boats, both for herring and for white fish. And the adoption of new forms of fishing was accompanied by a change in the social relationships which had formed around fishing; the old subjection in which the merchant had been a generally paternalistic and managing figure gave way to more purely commercial relationships with curers.

The change is sometimes taken to date from the disaster to the sixareen fleet in 1881 but the fact that replacement of boats so often led along new paths indicates that change had already occurred to precipitate the unconventional decisions made when the mass of the fishing population came to the point of replacing their boats. In the mid-seventies there were already scattered signs of a questioning of the old forms of fishing—of the subjection to the local curers and the use of the sixareen for the sole purpose of the distant haaf fishing. Firstly came a demonstration that big decked boats could successfully be used for great-line fishing. It was reported in the *Shetland Times* on 22 September 1877 that 'most of the fishermen have seen the Wick and Kirkwall boats which have been coming here for the last year or two, and cannot but be struck with the immense superiority of these over their own frail skiffs'.

[1] *Report on North Sea Fishing Industry, 1914*, App. II, p. 216.
[2] Ibid.

If they were so impressed it indicates a considerable shift in attitude. It was a deep conviction with Shetlanders that only the small boat, delicately handled, could be controlled with the sensitivity needed for the hailing of long-lines in rough water. This view was challenged when boats from Orkney and from the east coast of Scotland came to pursue the white fishing in Shetland waters. Already decked boats were being used with great success for great-line fishing from the mainland ports. In 1876 a handful of boats from Wick and the southern Moray Firth arrived in Shetland for the spring great-line fishing.[1] By local standards of catch their success was immediate and striking, and, since the shots were landed for local curing, the talk in the islands was soon of these successes. In 1877 there were six large boats engaged in the white fishing and they were averaging thirty to forty cwt. for a week's fishing; at this rate they accounted for the whole season's catch of a sixareen at the haaf in the space of two or three weeks.[2] Undoubtedly, then, the local people were being invited to question the fundamentals of their fishing method.

In the late seventies, preparations for large-scale change were being made in another way—experiments with large boats in herring fishing were beginning to show the results that would bring the herring fishing fleet in full force to the islands and which would offer a second justification for giving up the sixareens for the new large decked boats. Herring fishing in Shetland had been erratic and small in scale since the failures of the 1840s but a revival of interest came with the arrival of boats and curers from the south; some eighty-five boats attended for the herring fishing in 1876.[3] For two years the success was rather intermittent but it was enough to show that a considerable herring fishing was possible in Shetland during a season extending from May to September. In 1878 there came the first reasonably clear success in sustained herring fishing. Crews had been engaged at 20s. per cran, and by the end of August two boats had fished their complements of 200 crans.[4] This was enough to stir the interest. Local crews were now beginning to turn to herring fishing in decked boats of over 40-foot keel, which could also be used for spring white fishing, giving substantially higher yield than the traditional haaf. With curers prepared to invest in the new boats needed for this fishing, in 1879 some 200 boats were made ready for the herring fishing (some being incomers) and some sixareens

[1] FBR AF 29/9–10, Shetland, Weekly Reports, 10 June 1876.
[2] *Shetland Times,* 26 May 1877.
[3] FBR AF 29/9–10, Shetland, Weekly Reports, 10 June 1876.
[4] Ibid., 3 May, 24 Aug. 1878.

were added to the substantial number of new boats manned by local men.[1]

In retrospect, however, the late seventies appear as a halting preparation for the great surge that was to carry herring fishing in Shetland to its remarkable peak of 1884. The impetus was given from outside by both curers and crews from the mainland of Scotland; but such curers remained fairly few and the incoming boats scarcely increased in numbers beyond the levels of the first years of tentative exploration. Also limited, but of deep implication, were the adjustments in the life of the islanders themselves. The changes were not great but they irreversibly prepared the way for the over-turn of the old way of fishing and its place in general life of the landholders who comprised the great bulk of the population. Local curers began to look to the herring fishing from large boats as much as to the traditional ling fishing in sixareens. Furthermore, some fishermen experimented with the use of large decked boats in ling fishing and the number in local ownership showed a sudden jump from eight to twenty-eight between 1878 and 1879 and a further rise to sixty in the following year.[2] But the great majority still pursued the haaf in due succession after the sowing on the croft, and continued to engage with a local curer who would provide necessary supplies while taking the products of the haaf fishing. Some indeed had only recently acquired new sixareens for the haaf, and probably most were still persuaded that with the sixareen alone could the skills of great-line fishing be displayed; those who had crofts had been sternly advised that they should stick to the traditional system in all its details.[3]

The year 1880 saw a genuine over-turn of the old system. A fairly small fleet largely composed of local boats brought in hauls of herring that suddenly put Shetland in the forefront of interest for the many migrant Scottish crews. Some boats made over 600 crans in that season and the average for the whole fleet was 150 crans; if the large boats alone were considered, the average was considerably higher.[4]

The first result of the great fishings of 1880 was to bring curers from Scotland in much greater numbers and to build up swiftly the investment in herring fishing. It was not the first time that curers of the east coast had stretched out to create new branches of business in frontier areas largely devoid of commercial mechanism. Well entrenched in their home ports—with the capacity to manufacture barrels, with the

[1] *Shetland Times,* 6 Nov. 1880, 30 Sept. 1882.   [2] Ibid., 5 May 1882.
[3] Ibid., 6 Feb. 1878, 22 Mar. 1879, 10 May 1879.
[4] Ibid., 7 Aug. 1880, 1 Jan. 1881.

channels of credit open, their contacts with export markets working sensitively—they could swiftly create the physical means of curing at distant points and link the outlying areas with the main commercial network. Now they came in force to Shetland. One factor which made their way easy was the good harbour accommodation. They found considerable areas of sheltered water close enough to the main fishing grounds on all sides of the islands to serve as the multiple centres of a herring fishing, in which the boats would move between the grounds and harbours with good accommodation, every working day of the week. By leasing a piece of land in a suitable voe—space sufficent for one curing yard could be had for £10–£12 per annum—by erecting a jetty at the cost of about £200, and by assembling a cluster of huts for women workers and for the coopers, a curer quickly brought into being the physical structure of a curing yard; barrels were shipped to the site fully wrought and salt was landed to be stored in barrels.[1] When the workers had been transported to the site by boat, or perhaps when a few local workers had been inveigled into working the yard, all was ready for operation.

By 1884, 117 curing stations were laid out for the summer herring fishing of Shetland. A 'station' was the private possession of a single curer, usually comprising a jetty at the base of which was an open space where the stores could be laid out and where, often but not always completely exposed to the weather, the gutting and packing could proceed; nearby would be a cluster of temporary wooden huts in which the coopers and the yard staff would live. Unloading of fish was speeded up by the use of trolleys on which the herring were carried to the yard from the point of unloading. Some of the voes fairly bristled with the new jetties and the local life of the crofters with their small sixareens or fourareens was almost overwhelmed by the incoming boats and workers. In other cases, the stations were scattered in ones and twos along shorelines. In general, however, they fell into two main groups. Firstly, there were the stations which served the early fishing of May and June, taking place in the waters to the west and north of the islands. For this fishing, curing stations were established in the northerly isles and for some distance down the west coast of the island mainland; farthest north was Baltasound which had already emerged as the dominating centre of the early fishing. Then in July there started the distinct late fishing for which the boats had to strike south-east. From the very outset Lerwick was by far the outstanding centre in this fishing.

[1] Duthie, *Art of Fishcuring,* 27–31. Also, see map, p. 125.

The multiplying curing stations catered for a fleet of incoming boats which began to grow dramatically in 1882; by 1884, a peak year, the great majority of the 930 boats engaged came from the east coast of Scotland. But the response of the local crews to the new potential of herring fishing was also remarkable. Investment in large boats to be crewed by local men was the main basis of the expanding fishing up to 1881 and, while in 1882 and later the now numerous incoming boats accounted for the bulk of the catch, decked boats owned in Shetland and worked by Shetlanders continued their dramatic increase in numbers; by 1885 there were 342 of them.[1] Within the space of four years the long-established haaf fishing in open boats, which had seemed to be so deeply interwoven in the everyday life of the people, had been overshadowed by new forms of fishing. In this fishing decked boats were used for a cod and ling fishing in the spring, to be followed by a herring fishing starting in the northern Isles in May and moving to the more southerly locations centring on Lerwick in July. Not only did fishing incomes greatly increase but also the way of life of fishermen considerably changed, with some being drawn to a purely specialist pattern of activity.

When the Scotch curers engaged boats among the local population as well as from their own districts they introduced the Shetlander to a new scheme of social and economic relationship. The prices to be paid were established before the season and were fixed for its duration, having been determined by a process of bargaining in which the crew chose their employer as well as the curer choosing his crew; and when the season was over there might well be cash handed over to be spent freely where it would buy most. Crews were freed from bondage to one man and to one station. And to the opportunities in herring fishing there were added two other remunerative fishings. The spring fishing for cod and ling to be bought by the curers offered an employment for the larger boats which not only brought a further—and in this case fairly dependable—increment of income but also helped to meet the costs of the larger type of boat by allowing more continuous usage. Secondly, there was the winter fishing for haddock to be despatched fresh to the Scottish mainland, a fishing which had begun to rise in the late seventies. It was conducted in inshore waters both to the east and to the west, and the places of landing had to be accessible to the steamers which would take the fish to Aberdeen.[2] Thus it tended to be a fishing

---

[1] *Shetland Times*, 25 Sept. 1886.
[2] Smith, 'Trade in the Shetland Islands', 329; *Committee on North Sea Fishing Industry, 1914*, App. 10, p. 213.

of the central part of the islands, around Lerwick and Scalloway. Therefore, the basis existed for a year-round commercial fishing in which the larger boats would be used from March to September, first for line then for herring fishing, and the small boats during the winter for haddock fishing. Such exploitation could only be achieved within the central area of the islands and it was here that the specialist fishermen began to emerge.

The rapid re-equipment of the Shetland fishing population, so that it could to a varying degree take advantage of these new opportunities, depended on a quick and large investment in new boats. The cost of a boat and nets, which might well be of the order of £500, was generally beyond the immediate capacity of the partnerships of working fishermen; but with the big catches and high prices of the early eighties the ownership of a boat, which might be operated on the half-catch system with the owner taking a half share in profits and the crew the other half, was an attractive investment for landsmen. Shopkeepers and small merchants were drawn into investing in boats and supplied some of the needed funds.[1] Curers were interested not only in the profits of ownership but also in the right to acquire the catch of the boats' crews who were tied to them by debt; both the local curers, now breaking into herring fishing and the Scotch curers, who in fact bought most of the catch, invested in boats to be used by crews switching to the new modes of fishing.[2] The curers took both the share of the owner and the right to purchase the whole catch at rates probably lower than those ruling for the 'free' boats. Fishermen might hope to buy out the landsmen and become the full and untrammelled owners of their boats, although sometimes the landsman who had been original full owner might remain as a 'sleeping' shareholder in the enterprise.[3] In the boom years up to 1884 some crews did in fact secure this full ownership and it was in the central area of growing specialization that the fisherman who was also full owner was most common.[4] But the boom years were too few for anything like a full payment of debts and it is probable that when trouble came after 1884 only a few were free of a debt burden that now became crippling.[5]

A factor which added to this burden was the old dependence of the Shetland fisherman on advances by a capitalist of one type or another

[1] Smith, 'Trade in the Shetland Islands', 300.
[2] Commission on Crofters and Cottars, 1884, QQ. 19796, 19814.
[3] Smith, 'Trade in the Shetland Islands', 300.
[4] Ibid., 327; Committee on North Sea Fishing Industry, 1914, App. 10, p. 213.
[5] Shetland Times, 10 Apr. 1886.

to carry him through the season till he could repay out of the proceeds of the fishing.[1] Even prosperity did not end this need. The new providers of credit—the Scotch curers—became involved, as had been the local merchants before them and before that the lairds, in providing the means by which poor crofters could last out the season till the day of settlement. They were probably less interested than the merchants of the islands in the general trade that could be tied to the arrangements for fishing; but they were forced by circumstances to bring general stores to the more remote fishing stations and to give credit to the crews before the day of general settlement.[2] If the original negotiated price was high and the fishing turned out well, the crew would be more likely to clear the debt than they had ever been from ling fishing. But the season might still end in debt and the crew would then be bound to the same curer for the following season by bonds scarcely less tight than they had been over the haaf with the local merchants.[3] Debt meant that they would have to fish to the same curer over the following year and that the price obtained for their fish would be an imposed one below that ruling for the free boats.

In the period of the rapid re-equipment from 1880 to 1884 nearly all the fishing communities of Shetland acquired some boats of the new type and by 1886 only nine out of the forty-seven communities with some fishing population had no first-class boats.[4] But some patterns of strength and weakness were already evident. Most of the communities lacking large boats were set around the northwest rim, precisely the area from which the haaf fleet of small boats had mainly operated. On the other hand, the large boats were found in greatest numbers and in greatest density relative to fishing population over the middle sections of the east and west coasts of the mainland, with Lerwick far and away the greatest centre of boat ownership; some relation is here evident between growing specialization, with haddock fishing as an important element, and the readiness to equip to the fullest extent for the new large-boat fishings, both white and herring.

The year 1884 brought the whole herring fishing industry of Scotland to a crisis. The difficulties stemmed from over-production and were evident in a sudden sharp drop in the price of cured herring which was bound to affect every area. Thus, in Shetland as elsewhere, for the fishings of the next two years fishermen were offered much lower rates in their terms of engagement. Until 1887, however, many curers who

---

[1] *Commission on Crofters and Cottars, 1884*, QQ. 18548, 19271.
[2] Ibid., QQ. 18550, 19807–8.   [3] Ibid., QQ. 18603, 18896, 18899.
[4] FB Rep., *1886*, App. D.

were heavily in debt and nominally insolvent continued to be supported by the banks in the hope that a quick recovery of markets would allow them to repay debts. Thus the fleets which were fitted out for the summer fishings were little diminished in size and in Shetland at least record landings were made. But these same record fishings only made the chance of price recovery less likely; the financial returns to fishermen were poor, while many of the crews remained deep in debt. At last in 1887 the banks were beginning to foreclose on their debtors and with little credit offered for the new season the scale of fishing began to be curtailed. Thereafter till the mid-nineties the fleets which gathered for the Shetland fishing and the resulting landings remained well below the levels reached in the early eighties; but for the time being prices remained low.[1] Many of the Shetland fishermen had entered this period of recession in debt on account of their boats, and with the low prices the chances of their shaking clear were remote. Others with no boats of their own found it scarcely worthwhile to engage under half-catch terms on boats owned by landsmen. Correspondingly, the ownership of boats was no longer the attractive investment it had been. Altogether the stock of large boats began to fall and by 1895 there were only 267 on the register as compared with the 362 of 1886.[2] The shrinkage exposed more clearly the differences in the fishing strength of the different districts. In the northwest area most of the communities were found to be without any first-class boats and in the extreme south of Mainland also there was a marked thinning out; the central areas stood out as the most resistant to the difficulties of the time with remaining substantial investments in large boats.

By the late nineties prices began to rise again. Recovery brought the fleets to Shetland in greater numbers than ever and activity built up till 1,400 boats were assembling for the summer fishing.[3] After 1900 the powerful steam drifters were playing a big part in this fleet and landings soared to the peak year of 1905. But it was to be a peak rather than a plateau that the herring fishings had reached in this region and from 1908 to 1913 there was considerable fall in the Shetland herring fishings.

In the period of recovery Shetlanders responded much as they had done in the boom of the early eighties: that is, by quickly acquiring the boats needed for herring fishing. Funds were readily available. 'Any capable fisherman inclined for a share in a boat has no difficulty in getting outside partners—merchants and others.'[4] In 1905 there were

---

[1] FB Reps.   [2] FB Rep., *1895*, App. A, No. II.
[3] FB Reps.   [4] FB Rep., *1898*, 212.

over 400 first-class boats in the islands.[1] Until 1900 many of these boats were of the largest contemporary size, and in that year twenty-two had been fitted with steam capstans compared with only two in 1898.[2] A substantial proportion of the fishermen seemed to be keeping up with the latest trends in equipment and to be retaining or acquiring the position of owner. The rebuilding of the boat stock did not bring back the fishermen of the north-west coastline of the islands into the main stream of development; rather it was because of new acquisitions in the districts which had best resisted the decline of the late eighties that the total stock increased. The strength of the communities which could combine herring fishing with winter fishing for haddock was ever more clearly revealed.

When a boom which had been initiated with sailing vessels turned increasingly into a rush to invest in steam drifters the Shetlanders were left behind and by 1914 there were only four such vessels owned in the islands and only one owned even in part by fishermen.[3] The steamer, in fact, was difficult to fit into the rhythms of crofting since, to pay for itself, it had to be used at the East Anglian autumn fishing which impinged on a period when the crofter had to be looking to the land. But there was also by this time a number of full-time fishermen in Shetland who, equally with the crofter-fishermen, showed little inclination or capacity to acquire steamers.[4] One reason may have been financial. The crofters' closest financial links had been with local merchants and with the incoming curers, but these were not the men who played the largest part in the outside financing of the new fleet. It is easy to understand that fish salesmen, with their headquarters in east coast towns, would be loathe to acquire absorbing interests in boats partly owned, in peripheral areas, by men of whom they would have little direct knowledge; the very fact that the outside financiers' interests in the boats now went beyond those of the simple creditor had the effect of narrowing the geographical area within which funds were distributed for the aid of the fishermen. At such a time Shetland, like the west coast and Caithness, fell behind in the continual movement towards better means of fishing. But there was an additional reason for Shetland's neglecting the opportunities of the steam drifter. The basis of full-time fishing in Shetland was not so much the pursuit of herring throughout the year as the combination of herring fishing in the summer with line-fishing in the winter. The great-line spring fishing faded out

---

[1] FB Rep., *1905*, App. A, No. II.   [2] FB Reps., *1898*, 212; *1899*, 227.
[3] *Committee on North Sea Fishing Industry, 1914*, App. 19, p. 222.
[4] Ibid., App. 10, p. 213.

after 1905 but when small-line fishing for haddocks was set on a commercial basis in the islands the Shetlander found that he could make a living wholly out of fishing conducted from his own shores. And small-line fishing, implied the use of motor or sailboats rather than steamers.

Conclusion

IN THE first three-quarters of the nineteenth century, conditions—of markets, of technique, of yield from fishing grounds—favoured a general expansion of fishing round the Scottish coasts. But these conditions impinged differently upon different sections of a coast where the population was already generally committed to some form of fishing; this was not only because of variations in the local physical environment but also because social objectives tended to differ from one area to another. All regions—of the east and the west coasts respectively, of the northern isles, and of the Clyde area—responded in some degree by a form of expansion, but in doing so nevertheless remained very distinct from each other.

Along most of the east coast (apart from Caithness) in 1800 there were already to be found groups of men and communities fully dependent on fishing for a livelihood. Later, they showed a willingness to invest in new equipment, to seek new grounds, to develop newly profitable lines, and generally to alter their whole scheme of operations; the long-term result was the achievement of big increases in output and of a growing income to be shared by the ever greater number of families in the fishing population. But even here the ends of securing greater returns were not completely dominant. The desire to maintain a rough equality, the broad spread of property within the village and the strength of crew groupings based on kinship also determined the nature and efficiency of fishing operations.

In Caithness, fishing continued to be combined with the holding and working of land and this arrangement, which had been the basis of the first great expansion in the area, proved to be a tolerable long-run basis for herring fishing. Although equipment and yield never quite reached the levels found with the full-time fishermen from farther south, the numbers engaged in both farming and fishing held up well, at least till 1880, and the combination of the two occupations, in the general conditions of the area, seems to have provided a reasonably stable and acceptable standard of living. The social value of independence on the land was realized by the use of fishing as an auxiliary.

On the west coast also fishing was a widespread activity, almost always combined with the holding of some land. But in this area the

holding of land—a prior social aim—was itself attended by such difficulties (from the smallness of the holdings, the poverty of the soil and
the size of the rent) that capital and energy were drained away from a
fishing which remained subdued and ineffective. Indeed the west coast
men had to submit to a partial tutelage of east coast curers and crews
and it was as wage-earners rather than as independent owners that the
men drew most from fishing. This form of income played a considerable
part in allowing a large population to continue to exist on the basis of
small and inadequate holdings. In one west coast area, the Clyde estuary,
fishing gave a greater gain; and the possibility of earning a livelihood
by fishing alone was accompanied by, if it did not cause, a willingness
to give up the land, the working of which seemed to obstruct the full
realization of fishing gain; the numbers of crofter-fishermen diminished
to a tiny remnant collected around the upper part of Loch Fyne.

Land played still another rôle in Shetland, where expansion came in
the form of cod fishing—an activity for men temporarily deserting their
holdings—while the older ling fishing was left to be pursued as it had
been throughout most of the eighteenth century. It remained a seasonal
activity well integrated with the working of land, pursued with the full
efficiency possible for a small-boat fishing and built into a stable and
continuing system of livelihood.

Whatever the particular part played by fishing in a locality, involvement of a large proportion of the population at some level was almost
universal around the coasts, and the maximum in numbers of participants and in effect upon varying coastal communities was reached
about 1880.

Between 1880 and 1914 the rise in the national output of fish continued, at times more steeply than ever before; but in some areas
involvement was shrinking and the total number of those who could
in any sense be called fishermen diminished severely. Both herring and
white fish tended to be landed at fewer and fewer centres, mostly at this
stage by fishermen with highly expensive equipment. Thus the men in
the crowded sections of coast and the numerous communities in Banffshire, Moray and Aberdeenshire, together with the crews from Fife and
the few surviving communities south of the Forth, flourished in the
great surge in herring fishing between 1895 and 1913. But many of the
small villages, particularly in Caithness, in south Aberdeenshire and in
Kincardineshire, were forced into contraction and even extinction as
fishing communities, or they continued in the basis of the tenuous connection formed by acceptance of subordinate employment in boats
owned and worked elsewhere. Trawl fishing grew rapidly in Aberdeen

and in Granton even while many of the small communities fell into decline.

In Shetland, the first herring boom between 1880 and 1885 had involved the broad mass of the crofting population and had brought many crofter-fishermen into the position of owners of modern vessels for herring fishing; but the depression of the later eighties and early nineties severely cut down the number of men who had permanent possession of the means of up-to-date fishing and contracted the area in which the fishing population lived. When recovery came to the islands it failed to involve large areas where fishing had been a strong and long-continued tradition. Again, in the Clyde area the continuing success of full-time fishermen destroyed the last attempts to combine fishing with farming there.

On the west coast great numbers of crofter-fishermen were giving up the fishing side of their endeavour. But there still remained substantial groups, particularly in Lewis and Barra, who were able to raise their effort so as to obtain an increasing living from fishing while still holding on to their fragments of land. All in all, this time of expansion was shot through also with evidences of decline, particularly among the smaller communities, and was characterized by the very great shrinkage in the numbers of part-time fishermen—men who in the main had tried to combine fishing with farming.

By 1914, then, fishing had concentrated within solid industrial blocks with labour forces completely dependent on the one industry, with a lavish and expensive capital equipment (owned, in part by fishermen, in part by shore interests) and with complex systems of distribution and processing which maintained steady connection with a wide range of markets. Thus constituted, the industry—or industries—after being conditioned to expansion for over a century, had in the post-war period to suffer the chills of contracting or limited markets and of a natural environment in which good catches became less frequent.

The industry had now clearly separated into its two great branches—devoted to white and to herring fishing—and each had its particular difficulties. But, unquestionably, it was herring fishing which was deeper in the toils. The basic problem was one of markets. Some 80 per cent of the catch had in 1913 gone to Europe, almost entirely to Germany and Russia, with the latter becoming the greater market. When, therefore the new régime in Russia severely limited its imports, the Scottish export trade in herrings was irretrievably ruined. The German trade did eventually recover and in one year Soviet Russia made substantial purchases, but even in the twenties exports of herring never reached the pre-war average annual total. The thirties proved even more disastrous, particularly when Germany turned to policies of economic autarchy which included the building-up of German fishing capacity and a consequent exclusion of imports.

The fall in Scottish exports was desperately serious for the herring fishing industry. In the first place, there was little hope of selling elsewhere exports diverted from the lost markets. The domestic market for cured herring now scarcely existed and the sale of fresh herring, in its nature, could not be adjusted to the great fluctuations in supply which were inevitable in the herring trade. Thus, either failing foreign demand would bring a price collapse when the market took the full output of the industry as it had been running in 1914 or, alternatively, the restriction of sales necessary to stabilize price would cut severely the income generated by the industry. In fact, in the main, it was the second of these possibilities that came about; price did not collapse but less, much less, was sold.

The fall in income from herring fishing was serious because, in effect, large numbers of fishermen were equipped only for that type of enterprise. They used either sailing vessels to which paraffin motors had been added or, more importantly, steam drifters—a dependence which created its own peculiar problems. The addition of the paraffin motor did not turn the sailing vessel into a fully efficient fishing unit; in any case the sailing boats had all been built before 1907 and by the mid-twenties were at the end of their life. The steam drifter, on the other hand, was not easily adaptable to other purposes than herring fishing, the main alternative use for great-line fishing not being widely accepted as a possible means of earning a full livelihood. Furthermore, because of the heavy costs attending any use of the drifter, it had to be kept active for a large part of the year to have any chance of showing a surplus. Thus, to find a living, the fishermen had to use intensively a fleet which could only produce a commodity for which demand was severely limited. They could scarcely obtain a reasonable livelihood out of the income generated in the market to which they must sell and, almost as serious, they had no means of maintaining the existing capital stock or of replacing it by vessels more adapted to the needs of the times. Circumstances seemed to be forcing the herring fishing industry and its labour force into a slow attrition in which it would end up, after years of miserable earning, with no means of continuing the fishing.

In fact, the deterioration was well advanced by 1939. The conditions of successfully operating a steam drifter had proved to be beyond any reasonable management. The only inter-war period in which there was hope of covering costs, maintaining boats and replacing nets already outworn in the early twenties, was between 1925 and 1929; and even this period was interrupted by the severe problems of the coal strike of 1926. The early thirties were still worse, a fall in prices being coupled with low outputs and, until 1934, earnings fell to nominal levels. When some improvement in markets showed itself in 1935, the fleet had so far deteriorated as to be scarcely capable of taking advantage. Nearly all the vessels dated from before 1914 and many from 1906 or 1907, so that by the late twenties they were ageing seriously and by the mid-thirties desperately (many being by then thirty years old). There were no funds for replacement; and, especially when the rise in the price of steel after 1935 drastically increased the price of new vessels, even a new and fully efficient steam drifter could not be regarded as a hopeful investment. Even ordinary repairs and maintenance had to be neglected and by 1935 the position had been reached that catches were low and costs high because of the inefficiency of the craft. At this juncture, each year saw another few boats pass the point where it was no longer worthwhile to fit out for the fishing. The attrition then became obvious. The numbers in the fleet dropped seriously and the deterioration in gear was equally severe. Nets stiffened with age and, if lost, could not be replaced; thus boats had to operate with diminished drifts of sub-standard nets and they made catches considerably smaller than those achieved by fully equipped units. In general the total catch was kept down, in order to approximate to the amount that could be sold without serious drop in price; but the restriction on the quantity caught meant that the earnings on most boats were too low to allow the crews an adequate livelihood.

For many there was no way out of this vicious circle of deteriorating equipment and diminishing returns, and the diminution of the fleet in the later thirties is a sign that men were being forced, after years of desperate poverty, to give up fishing; they did not have the means to invest in preparing for types of fishing that would have been more remunerative. At the same time, even when unemployment was high, it became difficult to gather full crews, so miserable were the rewards. The only escape was by diversifying into other types of fishing. From the early twenties experiments were being made in the use of the Danish seine, a form of fishing in which a bag net was drawn through the water by a process of warping. It could well be operated from smallish motor-boats and make good catches of white fish in waters accessible to the stricken herring fishing communities. It might be expected, then, to provide the chance of diversifying into a branch of the industry in which the market was not subject to the same rigid limitations as existed for herring. Lossiemouth was the port most active in the new pursuit. No bonanza resulted, but reasonable returns were secured at low running costs. New ground was also being broken with the halting adoption of the diesel engine, which provided a means of power, had low running costs and did not suffer from the operational inefficiency of the paraffin motor. In fact, all through the thirties there were some who argued that a general-purpose boat, powered by the diesel engine, would provide a livelihood where steam drifters were hopelessly rigid in use and expensive to work; but, partly due to the shortage of funds for any replacements, these arguments made little impact and the chief form of diversification was the use of small boats equipped with paraffin motors and with either seine nets or, as of old, long-lines. The degree to which fishermen were able to make such subsidiary incomes from these forms of fishing varied from place to place; but in some communities, such as Buckie, there were certainly large numbers who found no other recourse than the failing herring fishing. The incomes from the small boats were no doubt merely auxiliary and offered no full escape from the dilemmas of herring fishing but they helped the fishing communities to survive through a time of great hardship. And they contributed to the gathering of capital for the time when genuine alternatives in the shape of the diesel boat became possible.

White fishing after 1919 was not faced with the same absolute shrinkage of markets as herring fishing. This section of the industry depended on the sale of fresh fish entirely in the home market and during the inter-war years per capita consumption of fish tended to increase in spite of a rise of fish prices that was greater than for prices generally. Yet, the Scottish, and in particular the Aberdeen, section of the trawl industry did not find it easy to draw advantage from growing markets. There were upward and downward swings in the scale of operations and in profits but, in the main, in the twenties and thirties it proved difficult for a fleet which was speedily brought back to its pre-war scale to sell its output with profit.

A major difficulty was the competition from the English ports and in particular from Hull. This Yorkshire port's fleet of long-distance trawlers, which made great catches in little-utilized distant grounds and was linked with an efficient marketing system, provided cheap but coarse fish and at the same time earned good profit. Such fish was not directly competitive with the Aberdeen product, which was of better quality and higher price, but the landings exercised a leverage on the total market and, together with the more directly competitive fish of Grimsby and Fleetwood, kept prices at levels that allowed the Aberdeen owners to make only intermittent profits.

This failure to make steady profits stemmed in part from the declining yields of the grounds to which Aberdeen, with its fleet mainly of short- and middle-distance trawlers, was committed. The North Sea, little fished during the war, gave bumper catches for the first two post-war years, but then began a severe diminution in the product of each day's fishing. The rise in running costs since pre-war days, estimated at 80 per cent, meant that the product of this diminished yield, even in conjunction with higher prices, failed to cover expenses in a majority of the inter-war years. The results were manifest in the poor profit record of Aberdeen compared with that of other trawling ports.

Low profits and frequent losses did not lead to the steady deterioration and ultimate shrinkage that we have seen in herring fishing. The size of the active fleet fluctuated, in part because, particularly in the early twenties, it was usual to keep boats ashore if they could not show profit (which led to some sharp contractions in the numbers at work), and in part because there was much buying and selling of second-hand vessels. Thus, in the early twenties, the registered fleet declined from its immediate post-war size, only to increase again between 1925 and 1929; then depression, manifesting itself in a severe drop in prices, brought another sharp contraction—which was followed, however, by a surge till 1936 when the fleet was greater than ever before. Rising numbers in Aberdeen were too often achieved by the buying of second-hand vessels and did not lead to technical regeneration. Indeed the average age of the vessels was always high, although not so seriously as in herring fishing, and it tended to increase as time passed.

The tenacity displayed by herring fishermen, who were also owners of boats, was matched in trawling by a policy which checked the rundown of the fleet even when owning groups were making no profit. The propensity of the owners, mainly non-fishermen, to keep activity going even when they were apparently earning no profit was partly the outcome of the diversity of their interests. They might make profit out of supplying the fleet or selling its product even when the fishing itself was nominally showing a loss, and the result might be an overall profit.

There were bright spots in the fishing industry of the inter-war period. The herring fishermen of the Clyde area came through with less strain than their counterparts in the east. The westerners had pioneered the use of motor boats powered by paraffin motors, which had lower operating costs and which proved efficient in the fishing conditions of the Clyde area. So equipped, the fishermen were less deeply hurt than the owners of steam drifters by fluctuations which might cut into the gross yield. Neither were their markets—mainly domestic—so depressed as those of the herring fishermen elsewhere. Furthermore, in the twenties they took advantage of the one form of export that flourished—the sale of herring uncured to German buyers who arrived with their own vessels to take away their purchases. Thus, against the trend elsewhere, the fishermen of the Clyde area enjoyed prosperity in the early twenties and came through the depression of the thirties in reasonably good order both as to equipment and as to income. Some Shetland fishermen, too, found a form of fishing that survived well through years of difficulty and depression. These were the specialized fishermen who, at least for part of the year, used motor boats in the expanding line-fishing for haddocks, to be sent by steamer to the greater market of Aberdeen.

The inter-war years also saw the completion of the trend first evident in the 1880s by which part-time fishermen and men in outlying and non-specialized communities give up any attempt at adding, even slightly, to their incomes by part-time fishing. Along the north-west mainland and among the inner isles the remaining recourse of the crofter-fisherman—the loch fishing for herring—was destroyed after 1931 by the disappearance of the shoals. Consequently the only widespread activity left was small-scale lobster fishing. On the Long Island it was the Minch fishing for herring that counted, and for a period of years after 1919 the sailing vessels with their auxiliary motors and low working costs seemed to survive and pay as well as did the steam drifters; but, in the end, the problem of age and obsolescence, becoming acute in the thirties, largely destroyed the fleet and left in the islands only a tiny remnant of men making anything from fishing. The wooden sailing vessels nearly all dated from before the time when the demand for steam drifters had risen so greatly on the east coast—that is, from before 1907. When inevitably they became too old for use, there were neither funds for replacement nor any very obvious type of craft to take their place. In Shetland too the remaining crofter-fishermen, already greatly diminished in numbers by 1914, struggled along with a fleet of outdated sailing boats, often partly modernized by addition of paraffin engines; but they too succumbed in the end to the problems of obsolescence and depression.

By 1939 Scotland's fishing industry was much shrunken from the expansive days of the 1880s and from the narrower prosperity of the years just before 1914. In places there had been complete collapse. Instead of the once active fishing there might now be found only decaying harbours, shells of villages which, where they survived, turned to other purposes, and farms where the tenants had no thought for the sea. Even at the still active centres, the basis of fishing had been eroded by the drift of labour to other occupations and by the dissipation of capital in the years of hardship. Yet, on the whole, this still remained a major industry, resilient and capable of adaptation and possibility of expansion. Many families had an indestructible attachment to the trade and each succeeding generation still produced its recruits for the sea in an unbroken tradition of centuries. The newer trawling communities, if they lacked such long-standing traditions, yet had much skill and experience, particularly among the skippers. The financial structure also was tough and besides the involvement of many of the fishermen, a complex web linked diverse interests to the continuance of fishing. Out of such enduring aspects came a capacity for renewal which was to show itself after 1945.

Bibliography

MANUSCRIPT SOURCES

British Fisheries Society Records. The papers used in this book are housed in the Scottish Record Office under the classification GD9.

Fishery Board Records. The body which administered the Scottish fisheries was successively: The Commission for the Herring Fishery (1809–41) (Scotland) (1842–81), The Fishery Board for Scotland (1882–1914). The Reports of these bodies form a continuous series and are couched in forms which did not change with changes of nomenclature and detailed administrative arrangement. For convenience, then, the single term 'Fishery Board' is used throughout, both in text and in footnotes.

Seaforth Muniments.

Board of Trade Register of Companies.

NEWSPAPERS

Aberdeen Free Press
Banffshire Journal
Berwickshire News
Daily Free Press
Fife Herald
Fife News
Fishing News

Fish Trades Gazette
Fraserburgh Advertiser
John O'Groat Journal
Northern Ensign
Peterhead Sentinel
Shetland Times

OFFICIAL REPORTS

Committee appointed to inquire into the State of the British Fisheries, 1785, Reports from the House of Commons, 1803, X.

Committee appointed to inquire into the State of the British Herring Fisheries, 1798, Reports from the House of Commons, 1803, X.

First Report from the Select Committee appointed to inquire into the Laws relating to the Salt Duties, 1801, III.

Report from the Select Committee appointed to take into consideration the Laws relating to the Salt Duties, 1818, V.

Report from the Select Committee appointed to inquire into the state of the Circulation of Promissory Notes under the value of £5 in Scotland and Ireland, 1826–7, VI.

Report from the Select Committee appointed to inquire into the condition of the Population of the Highlands and Islands of Scotland, and into the practicability of affording the people relief by means of Emigration, 1841, VI.

Report from Her Majesty's Commissioners for inquiry into the Administration and Practical Operation of the Poor Law in Scotland, 1844, XX–XXVI.

Extract from Captain Washington's (unfinished) Report on the Damage caused to Fishing Boats by the Gale of 19th August, 1848, 1849, LI.

Report to the Board of Supervision, by Sir John M'Neill, G.C.B., on the Western Highlands and Islands, 1851, XXVI.

Reports on and since the year 1848 on the subject of the Fishery Board in Scotland, 1856, LIX.

Reports addressed to the Lords Commissioners of Her Majesty's Treasury in 1856, on the subject of the Fishery Board in Scotland, 1857, XV.

Report of the Commissioners appointed to inquire into sea fisheries of the United Kingdom, 1866, XVII–XVIII.

Reports of the Select Committee appointed to inquire into the policy of making further grants of public money for the improvement and extension of harbours of refuge, 1857, XIV; 1857–8, XVII.

Report of the Commissioners appointed to complete the inquiry into the terms recommended in the Report of the Select Committee of the House of Commons in 1858, on Harbours of Refuge, 1859, X.

Fourth Report of the Commissioners on the employment of children, young persons and women in Agriculture, 1870, XIII.

Second Report of the Commissioners appointed to inquire into the Truck System (Shetland), 1872, XXXV.

Report on the Herring Fisheries of Scotland by the Inspector of Salmon Fisheries for England and Wales and the Commissioners of Scotch Salmon Fisheries, 1878, XXI.

Report from the Select Committee appointed to inquire into the expediency of continuing the present system of Branding Herrings, and into the Appropriation of the Revenue raised from the Brand Fee, 1881, IX.

Report of the Select Committee on Harbour Accommodation, 1883, XIV; 1884, XII.

Reports of the Commissioners of Inquiry into the Condition of the Crofters and Cottars in the Highlands of Scotland, 1884, XXXII–XXXVI.

Report of the Royal Commission on Trawl Net and Beam Trawl Fishing, 1884–5, XVI.

Report by a Deputation to the Continent to inquire into the new Branding Regulations, 1890–1, LXII.

Report from the Select Committee on Sea Fisheries, 1893–4, XV.

Report of the Departmental Committee on the Sea Fisheries of Sutherland and Caithness, 1905, XIII.

Report of the Scottish Departmental Committee on the Scottish Sea Fishing Industry, 1914, XXXI.

BOOKS AND ARTICLES

ANDERSON, JAMES General View of the Agriculture and Rural Economy of the County of Aberdeen (Edinburgh, 1794).

ANSON, PETER F. Fishing Boats and Fisher Folk on the East Coast of Scotland (London, 1930).

ARBUTHNOT, JAMES An Historical Account of Peterhead, from the Earliest Period to the Present Day (Aberdeen, 1815).

BARRY, GEORGE The History of the Orkney Islands (Edinburgh, 1805).

BERTRAM, JAMES G. The Harvest of the Sea (London, 1869).

BUCHAN, PETER Annals of Peterhead (Peterhead, 1819).

BREMNER, DAVID The Industries of Scotland (Edinburgh, 1859).

CARTER, G. 'The Fishing Industry', in Donald Omand, ed., The Caithness Book (Inverness, 1972).

DE CAUX, J. W. The Herring and the Herring Fishes (London, 1881).

CLARK, W. FORDYCE The Story of Shetland (Edinburgh, 1906).

COWIE, ROBERT Shetland: Descriptive and Historical (Edinburgh, 1874).

COULL, JAMES R. 'Fisheries in the North-East of Scotland before 1800', Scottish Studies, 13, 1969.

CRANNA, JOHN Fraserburgh: Past and Present (Aberdeen, 1914).

DAY, J. P. Public Administration in the Highlands and Islands of Scotland (London, 1918).

DONALDSON, JAMES General View of the Agriculture of the County of Banff (Edinburgh, 1794).

DUFF, R. W. 'The herring fisheries of Scotland', Fisheries Exhibition Literature (London, 1884), Vol. VI.

DUNCAN, JOE 'Capitalism and the Scots Fisherman', The Socialist Review, Vol. II (XI), 1909.

DUTHIE, R. J. The Art of Fishcuring (Aberdeen, 1911).

EDMONDSTON, ARTHUR A View of the Ancient and Present State of the Zetland Islands, 2 vols. (Edinburgh, 1809).

FALL, R. Observations on the Report of the Committee of the House of Commons appointed to Inquire into the State of the British Fishery (Dunbar, 1786).

FEA, JAMES Present State of the Orkney Islands Considered (Edinburgh, 1884).

FRASER, ROBERT A Review of the Domestic Fisheries of Great Britain and Ireland (Edinburgh, 1818).

GIFFORD, THOMAS Historical Description of the Zetland Islands (London, 1786).

GOODLAND, C. A. Shetland Fishing Saga (Lerwick, 1971).

GRAHAM, MICHAEL (ed.) Sea Fisheries and their Investigation in the United Kingdom (London, 1956).

GOURLEY, GEORGE Fisher Life; or, the Memorials of Cellardyke and the Fife Coast (Cupar, 1879).

GRAY, MALCOLM 'Fishing Villages, 1750–1880', The North-East of Scotland. A Survey prepared for the Aberdeen

Meeting of the British Association for the Advancement of Science, 1963 (Aberdeen, 1963).

GRAY, MALCOLM 'Organisation and growth in the east coast Herring fishing', in P. L. Payne, ed., *Studies in British Business History* (London, 1967).

HALCROW, CAPT. A. *The Sail Fishermen of Shetland* (Shetland, 1950).

HENDERSON, JOHN *General View of the Agriculture of the County of Caithness* (London, 1812).

HENDERSON, JOHN *General View of the Agriculture of the County of Sutherland* (London, 1812).

HERBERT, DAVID (ed.) *Fish and Fisheries, A Selection of Prize Essays of the International Fisheries Exhibition* (Edinburgh and London, 1883).

HEADRICK, JAMES *General View of the Agriculture of the County of Angus or Forfarshire* (Edinburgh, 1813).

HIBBERT, SAMUEL *A Description of the Shetland Islands, comprising an Account of their Geology, Scenery, Antiquities and Superstitions* (Edinburgh, 1822).

JAMIESON, PETER *Letters on Shetland* (Edinburgh, 1849).

JOHNSTONE, JAMES *British Fisheries, their Administration and their Problems* (London, 1905).

KEITH, GEORGE SKENE *A General View of the Agriculture of Aberdeenshire* (Aberdeen, 1811).

KEMP, JOHN *Observations on the Island of Shetland and their Inhabitants* (Edinburgh, 1801).

KNOX, JOHN *Observations on the Northern Fisheries* (London, 1786).

KNOX, JOHN *A Tour through the Highlands of Scotland and the Hebride Isles, in 1786* (London, 1787).

KNOX, JOHN *A View of the British Empire, more especially Scotland* (London, 1784).

LEVI, LEONE *The Economic Condition of Fishermen* (London, 1883).

LEWIS, SAMUEL *Topographical Dictionary of Scotland*, 2 vols. (London, 1847).

LINDSAY, PATRICK *The Interest of Scotland* (Edinburgh, 1733).

LOCH, DAVID *Essays on the Trade, Commerce, Manufactures and Fisheries of Scotland*, 3 vols. (Edinburgh, 1778–9).

MOFFATT, WILLIAM *Shetland: the Isles of the Nightless Summer* (London, 1934).

MACCULLOCH, LEWIS *Observations on the Herring Fisheries upon the North and East Coasts of Scotland, etc.* (London, 1788).

MCINTOCH, WILLIAM CARMICHAEL *The Resources of the Sea* (Cambridge, 1921).

MILLER, JAMES *The History of Dunbar* (Dunbar, 1830).

MILN, W. S. *An Exposure of the Position of the Scotch Herring Trade in 1885* (London, 1886).

MILN, W. S. 'The Scotch East Coast Herring Fishing', *Fisheries Exhibition Literature* (London, 1884), Vol. XI.

MARCH, EDGAR J.   *Sailing Drifters* (London, 1952).
MATHER, J. Y.   'Aspects of the linguistic geography of Scotland: III fishing communities of the east coast (Part 1)', *Scottish Studies*, xiii, 1969.
MITCHELL, JOHN   *The Herring. Its Natural History and National Importance* (Edinburgh, 1864).
A NORTH BRITON   *General Remarks on the British Fisheries* (London, 1784).
NEILL, PATRICK   *A Tour through some of the Islands of Orkney and Shetland* (Edinburgh, 1806).
New Statistical Account of Scotland, 15 Vols. (Edinburgh, 1835–45).
O'DELL, ANDREW C. *The Historical Geography of the Shetland Islands* (Lerwick, 1939).
Ordnance Gazetteer of Scotland (Edinburgh, 1906).
PITCAIRNE, GEORGE   *A Retrospective View of the Scots Fisheries* (Edinburgh, 1787).
PENNANT, THOMAS   *Tour in Scotland and Voyage to the Hebrides, MDCCLXXII* (Chester, 1774).
PLOYEN, CHRISTIAN *Reminiscences of a Voyage to Shetland, Orkney and Scotland* (Lerwick, 1894).
RAMPINI, CHARLES   *Shetland and the Shetlanders* (Kirkwall, 1884).
RICHARDS, ERIC   *The Leviathan of Wealth* (London, 1972).
SHERRIFF, JOHN   *General View of the Agriculture of the Shetland Islands* (Edinburgh, 1814).
Statistical Account of Scotland, 21 vols. (Edinburgh, 1790–8).
STEPHENSON LOCOMOTIVE SOCIETY   *The Highland Railway Company, 1855–1955* (London, 1955).
THOM, WALTER   *The History of Aberdeen* (Aberdeen, 1911).
THOMSON, JAMES   *The Value and Importance of the Scottish Fisheries* (London, 1849).
THOMSON, JAMES   *On the Existing State of our Herring Fishery* (Aberdeen, 1854).
Topographical and Historical Gazetteer (Glasgow, 1842).
TUDOR, JOHN R.   *The Orkneys and Shetlands: their Past and Present State* (London, 1883).
SOUTER, DAVID   *General View of the Agriculture of the County of Banff* (Edinburgh, 1812).
WHITE, P.   *Observations on the Present State of the Scotch Fisheries* (Edinburgh, 1791).
WILSON, JAMES   *A Voyage round the Coasts of Scotland and the Isles*, 2 vols. (Edinburgh, 1842).

Index

Aberdeen, 149, 158, 211–12, 215; advantages as trawling port, 167, 170–1, 215; as market, 43, 162, 167, 170, 217; capital for development of trawling, 178–9; communications, 170, 172; curing facilities, 169, 172–3; employment, 174–5; fish market, 171–2; fleet of trawlers, 167, 170, 173–4, 215–16; harbour development, 76; herring fishing, 78, 85, 152; increase in size of vessels, 173; landings of white fish, 166, 168, 170, 171; market facilities, 167, 169, 171, 176; migration into, 169, 175–6; overloading of facilities, 172; ownership of steam drifters, 157; prices, 172, 173–4, 176; role of curers, 167, 172; trawlowners, 178–80, 216
Aberdeen Bay, 167
Aberdeenshire, 45; boat design, 46; investment in boats, 22, 47; rise of herring fishing, 38–9; villages, 12, 211
Anstruther, 75, 78–9, 91, 157, 164
Arbroath, 15, 76
Ardrishaig, 197, 199
Argyllshire, 181
Arran, 120, 197
Atlantic, 1–2, 4, 114, 171
Auchmithie, 42, 90
Auction system, 65, 93, 148, 151–2; see also Fife; Peterhead
Austria, 60
Ayrshire, 4

Badachro, 191
Balaclava Breakwater, 77
Ballantrae, 197
Baltic market, 52
Baltic Sea, 1–2
Banff, 157
Banffshire, 45, 47; control by landlords, 21–2; family labour, 13; share systems, 21–2; villages, 10, 12, 211
Banks, 155
Barra, 86–7, 112, 114, 117, 182, 191–3, 195–6, 212
Barra fishery district, 195

Belgium, 60
Bervie, 43
Berwick, 14
Berwickshire, 22–3, 36, 39, 48, 84–5
Billingsgate, 122, 170
Birmingham, 9, 16
Boat design, 34–5, 46, 82–3; see also Caithness; Loch Fyne fishery; Moray Firth coast; Shetland
Boat ownership, 13, 98–9, 158, 163, 165, 194–5; see also Caithness; Moray Firth coast; Peterhead; Shetland
Boats, cost, 21, 35, 46, 97–8, 107, 132, 187, 193, 198; size, 15–16, 32, 35, 46, 82, 84, 104, 106–7, 111, 115–16, 120, 126, 178, 183–4, 188–9, 198, 208
Boddam, 44–5, 75
Bounties, 5–6, 53, 58
British Fisheries Society, 5–6, 33, 77
Broadsea, 10
Bruce of Sumburgh, 132
Buchan (north-east Aberdeenshire), cod fishing, 17; curing of haddocks, 17; engagements by curers, 40; exports to Continent, 51, 56; herring fishing, 39; seasonal migration, 48
Buckhaven, 14, 164
Buckie, 10, 14, 46, 75, 78–9, 91, 152–3, 157–8, 165, 215
Burghead, 75
Burntisland, 27, 64
Burray, 127
Buss, 5–6, 32, 105, 120
Bute, 120

Caithness, 40, 48, 55, 81, 84, 88, 106–7, 109, 164, 191, 208, 210–11; engagement system, 30; farming and fishing, 31, 36, 50, 81, 92, 210–11; fishing stations, 33–4, 73, 79; growth of population, 36, 80–1; indebtedness, 81, 99; influx of hired hands, 108; markets, 19, 29, 30; ownership of boats, 34, 38, 98; rise of herring fishing, 29–38; scale of herring fishing, 33, 35, 39, 85; type of boat, 46
Caledonian Railway Company, 172

Campbeltown, 51, 197, 199
Campbeltown fishery district, 186, 190
Carradale, 197
Cellardyke, 46
Cluny Harbour, 75-6
Clyth, 14
Cod and ling fishing, fishing grounds,
2-3, 114; markets, 18, 26, 44, 90-1,
115; methods, 17, 114-15; see also
Buchan; Fife; Fraserburgh; Moray
Firth coast; Shetland cod fishing
Cod curing, 44, 138; by curers, 26, 115,
117; by fishermen, 18, 115, 117;
methods, 18, 91, 115, 169, 172-3
Collieston, 44
Communications, 4, 170, 182-3
Continental market, 51-3, 59, 61-2,
110, 131, 139, 147, 149, 193, 212-13;
development by Dutch, 52, 60;
organization, 56, 66-7; see also
Buchan; Moray Firth coast; Wick
Congested Districts Board, 182, 191
Cove, 43
Crail, 164
Cruden, 12
Cullen, 12
Curers, debts, 71-2; development of
Caithness fishing, 29-30; develop-
ment of Minch fishing by east coast
firms, 110, 152, 211; development of
Shetland fishing by east coast firms,
64, 87, 153; investment in herring
fishing, 151; loans to fishermen, 47,
69, 99; origins, 30, 37, 50, 62; profits
and losses, 30-1, 37, 56, 69-72, 146;
sources of funds, 66-9, 99; see also
Minch fishing; Shetland herring
fishing

Danish seine net, 214
Danzig, 55
Davis Straits, 138
Diesel engine, 215
Dodds Steam Fishing Company, 179-
80
Dodds, W. H., 180
Dogger Bank, 171
Dunbar, 18, 75-6, 78
Duncansby Head, 14
Dundee, 90, 172

Earnings, 49, 93, 109, 111, 124-5, 127,
177-8; cod and ling fishing, 19, 115;
haddock fishing, 43-4, 97; herring
fishing, 41-2, 47, 87, 93-6, 106-8,

112-13, 121-2, 146, 184-5, 187, 192,
196, 213; see also haaf fishing; hired
hands; Loch Fyne fishery; Peter-
head; Shetland
East Anglia, 149, 159, 164; migration
from Scotland, 87-8, 92, 153-4, 196
East Lothian, 84
East Neuk of Fife, 9-10, 75, 78, 164
Edderachylis, 107
Edinburgh, as market, 19, 26, 43-4, 57
Engagement system, 56, 65, 68-9, 71,
87, 107, 110, 112, 147, 163, 193, 198,
204, 206; decline, 148; used to
establish herring fishing, 30-1; see
also Buchan; Caithness; Shetland,
Shetland herring fishing
England, 87, 166, 179; as market, 89,
172, 183
Eyemouth, 17-18, 78, 88-90, 92, 98, 164
Europe, 2, 5, 7, 56

Falls of Dunbar, 29
Family labour, 48-50, 92, 189-90
Farming and fishing, 14-15, 119, 123,
126, 132, 182, 199, 211-12, 217; in
North West, 7, 101-3, 107, 109, 111,
115, 117, 181, 187, 196-7, 211-12;
see also Caithness; Moray Firth
coast; Loch Fyne fishery; Shetland
Faroe Islands, 171
Faroe fishing, 138
Fetteresso, 43
Fife, 45-7, 79, 88; auction system, 65;
cod fishing, 17, 26, 91; dependence
on Anstruther Union Harbour, 164;
haddock fishing, 16, 43, 88, 90;
maintenance of equality, 164; markets, 43; migration to East Anglia,
87, 92, 153, 156; migration to northern fishings, 39, 48, 75; rise and
decline of herring fishing, 5, 78, 88;
share system, 98; villages, 211
Fifie class of boat, 84
Findochty, 45
Findon, 17, 42-3, 88
Firth of Clyde, 51, 57, 101, 105, 118,
120, 122, 151, 197-8, 200-1, 212-13,
216-17
Firth of Forth, 15, 18-20, 27, 29, 37, 57,
64, 146, 211
Fish salesmen, arrangements with
fishermen, 155-6, 208; as source of
credit, 156; control of fishing, 156;
functions as managers, 157; growing
wealth, 168; investment in trawling,

Fish salesmen—*contd.*
180; many interests, 156; origins, 178; role in trawling industry, 168; *see also* Fraserburgh
Fishery Board, 74–5, 77–8, 98, 182; control of herring fishing, 53–4, 57, 65; loans for boats, 182
Float-drave, 19–20
Footdee, 166, 175
Forbes, 17
Forfar, 43
Forth and Clyde Canal, 27
Fourareen, 129
France, 60
Fraserburgh, 12, 72, 79, 82, 157, 165; cod fishing, 91; composition of crews, 22; harbour development, 76–7; investment in boats, 82–3; number of curers, 64, 71; ownership of steam drifters, 157–8; population, 10; position of salesmen, 157; scale of fishing, 39, 78, 82, 152; share system, 23

Gairloch, 114, 116, 191
German import firms, 66–7, 69, 71, 99, 217
Germany, 53, 55, 61, 110, 131, 147, 149, 173, 212–13
Glasgow, 57, 183; as market, 43, 50, 89, 100, 119, 122, 169, 186, 199
Gordon of Cluny, 76
Granton, 149, 162, 167, 212
Great-line fishing, 44, 79, 90, 116–17, 124, 154, 190–2, 196–7; earnings, 44, 115–16, 188–9; method of operation, 17–18, 114–15; *see also* Shetland great-line fishing
Greenock, 30, 50, 122
Grimsby, 166, 169, 171–2, 216
Ground-drave, 18

Haaf fishing, based on fishing tenures, 132, 134, 142; boats, 129, 132, 140; catches, 129; curing, 130; decline, 204; determination of prices, 133–4, 142; earnings, 134; fishing grounds, 127–30; fishing stations, 128–30; lines, 129; make-up of crews, 132; merchants take control, 136, 140–1; mode of operation, 129–30; part played by landlords, 128, 132–4; season of fishing, 128; share system, 132–3; supply of credit, 132; yields, 134

Haddock curing, household system of production, 43–4, 88–90, 169; methods, 17, 43–4, 169; yard production, 88–9; *see also* Buchan; Moray Firth coast
Haddock fishing, 1, 16, 79, 127, 154; catches, 42–3; engagement terms, 49, 167; fishing grounds, 3; markets, 42, 49, 88–9, 169; method of capture, 15–16, 90; season, 15, 92; *see also* Buchan; Fife; Fraserburgh; Shetland haddock fishing
Half-dealsmen, 45
Hamburg, 55, 61
Harbours, 3–4; development, 73, 79, 152, 182, 189; lacking on east coast, 15, 24; lacking in north west, 189; natural facilities, 3–4; *see also* Aberdeen; Fraserburgh; Moray Firth coast; Wick
Harbours and Passing Tolls Act, 1861, 76
Harris, 103
Hay and Ogilvy, 139
Hebrides, 114
Helmsdale, 36, 75, 81, 165
Herring brand, 54, 61, 66–7
Herring curing, Dutch example, 5, 52; labour employed, 36–7, 53–4, 65; method, 31, 53–5, 65, 151; reddening of herring, 29; size of firms, 37, 64, 71
Herring fishing, 21, 27, 213; catches, 4, 42, 95, 184, 186–7; combined with white fishing, 42; crisis of 1884, 146–8, 193; geographical distribution, 1–2, 84–6; growth of output, 58–9, 63, 82, 146; loch fishing, 104–6, 183–8, 217; method of fishing, 18–21; predominance over other fishings, 149, 154; winter fishing, 84, 88; *see also* Minch fishing; Shetland herring fishing
Hired hands, 45, 48, 175–6, 200; earnings, 45, 95–6, 109, 114, 193, 200; from west coast, 36–7, 108, 113–14
Holland, 60–1; herring fishing, 5–6, 9, 52–3
Holy Island, 17
Hull, 166, 171–2, 215

Ice, imports from Norway, 173; supplies, 173
Iceland, 138, 173
Inner Hebrides, 4

Investment, 184, 187; by fishermen, 45–7, 69, 98–100, 110, 113, 120, 158, 165, 204–5; see also Aberdeenshire, curers, fish salesmen, Fraserburgh, Shetland herring fishing, villages
Ireland, 51–2, 59, 139, 193
Isle of May, 17

Johnshaven, 14, 175–6
Johnstone and Sherrett, 178
Johnstone, Peter, 169

Kilbrennan Sound, 120, 197
Kincardineshire, 39, 46, 48, 79, 162, 211
Kintyre, 4, 101, 118, 123, 197
Kirkwall, 127, 200
Kyle of Lochalsh, 183, 186, 191

Latheron, 34, 36
Leith, 19, 26
Lerwick, 87–8, 203, 205–6
Lewis, 87, 110, 115–17, 147, 189, 191–2, 195–6, 212; curing stations, 86–7, 112; loans for boats, 182
Ling, see cod and ling
Lobster fishing, 124, 197, 217
Lochboisdale, 197–8
Lochbroom, 104
Lochbroom fishery district, 186, 188, 190, 195
Loch Carron and Skye Fishery District, 106, 185–6, 188
Loch Fyne fishery, catches, 119, 199, 200; competition between drift net and ring net, 120–2, 197–9; earnings, 121–2, 211; farming and fishing, 123, 197, 199, 211–12; fishing grounds, 198–9; marketing, 57, 119, 120–2; nets and gear, 120–2; ownership of boats, 119, 123, 198, 200; price advantages, 119, 186, 197; share system, 119, 123, 200; type of boat, 119, 120, 122, 184, 187–8, 198–9, 200, 216–17
Lochgilphead, 197, 199
Loch Goil, 119
Loch Long, 119
London, 18, 30, 43, 44, 50, 66, 172, 186
Long Island, 113, 152, 195, 217
Lossiemouth, 76, 165, 215
Lothians, 20, 36, 39, 48, 84, 92
Lowestoft, 153
Lybster, 34, 81, 164

McCombie, James, 71
Macduff, 75, 85

Mallaig, 183
Mar Bank, 17
Markets, 131, 183, 215; for cod and ling, 131; for haddock, 88–9, 204; for herring, 19, 29, 30, 32, 37, 50, 51–3, 56–7, 87, 110, 120–2, 139, 147, 149, 150, 213; see also Continental market
Marketing, 190–1
Marriage customs, 13–14, 24, 48, 80–1; see also Buchan
Mediterranean, 44
Meff, William, 168, 178
Meff Brothers, 157, 169, 178
Merchants, in North West, 107, 112–13, 115–16, 189; in Shetland, 134, 138, 140–5
Methuen, James, 64
Migration, 35, 163; of crews, 32, 36–7, 39, 40–1, 48, 85–7, 92, 120, 153, 163–4, 196–7, 201; of hired hands, 36–7; of women, 36–7; see also Buchan, Fife, Moray Firth coast
Minch, 114, 171
Minch fishing, 96, 110–14; catches, 113; curing stations, 86, 192; influx of east coast boats, 152; local participation, 112, 195; rise of summer fishing, 86–7, 110, 152; scale, 87, 152–3, 188, 193, 195, 197, 217; winter fishing, 152–3
Montrose, 9, 85, 91, 155, 163
Moray Firth coast, 108; boat design, 46; closed to trawling, 171; cod fishing, 17; curing of haddocks, 17; engagement by curers, 40; exports to Continent, 51, 56; farming and fishing, 14; haddock fishing, 9; improvement of harbours, 79; large boats used in Shetland great-line fishing, 201; ownership of boats, 98; rise of herring fishing, 39; scale of herring fishing, 39, 85; seasonal migration, 32; villages, 38
Motor boats, 184, 199, 213, 217
Mull, 101

Nairn, 154, 165
Napier Commission, 181
Ness, 182, 189
Nether Buckie, 76
Netmaking, 48
Nets, number in drift, 32, 35, 46, 82, 98, 104, 106, 120, 122, 149, 155, 184; use of cotton nets, 82

Newhaven, 26
North British Railway Company, 170, 172
North of Scotland Bank, 71–2
North Ronaldsay, 126
North Sea, 3, 52, 56, 59, 171, 216
North Shields, 169
Norway, 1–2; competition with Scotland, 56, 60–1, 62–3, 67; supply of ice, 173

Oldcastle, 176
Orkney, 124–7, 171, 201
Orphir, 124
Otter Point, 121
Otter trawl, 174
Outer Hebrides, 3–4, 182, 195
Outer Isles fishing district, 186, 190

Perth, 43
Peterhead, 41, 71, 84, 91, 154, 158, 163, 165, 170, 176; concentration of herring fishing, 152; development of harbour, 74, 76–7; earnings on steam drifters, 161–2; engagements by curers, 40; new auction system, 148; ownership of steam drifters, 157; place of steam drifters, 158; rise of herring fishing, 39; scale of herring fishing, 78, 152
Poland, 149
Port-Errol, 75
Portessie, 45
Portgordon, 12
Portknockie, 12, 45
Port Seton, 164
Portugal, 60
Prices, 173–4, 176, 182–3, 215–16; of cod and ling, 134–5, 144, 190–2; of haddock, 89, 172; of herring, 30, 41, 47, 50, 56–8, 62–3, 70, 72, 82, 119 147–8, 185–6, 193, 197, 207, 212–14
Public Works Commissioners, 76
Pulteneytown, 33, 35, 81
Pyper, William, 178

Railways, 170, 172, 183
Rathven, 18
Red Star Sea Fishing Company, 179
Rhine, 52, 60
Rhineland, 60–1
Ross, 111
Rousay, 124
Russia, 60, 212

Saint Monance, 10, 164
Salt Laws, 6
Scalloway, 205
Scandinavia, 14
Scarborough, 169
Scotland, markets within, 172
Share systems, 21–4, 44–8, 67, 98–9, 104, 111, 119, 123–4, 159, 200; see also Banff; Fife; Fraserburgh; Loch Fyne fishery
Shetland, 171, 181; boat design, 129, 136–7; cod fishing, 136–8, 211; credit, 144–5; determination of prices, 142; earnings, 135; engagement systems, 143; farming and fishing, 130–2, 134–5, 140, 142, 202, 211, 217; fishing grounds, 1–2, 4; fishing tenures, 132, 142; gains by landlords, 134–5; indebtedness of fishermen, 133, 144–5, 204–7; inshore fishing, 127; landownership, 131, 134–5; ownership of boats, 124, 132, 139, 144, 204–5, 207, 212; purchase of land by merchants, 140, 142; role of merchants, 136–8, 140–5
Shetland great-line fishing, 200–1, 204, 208–9; incoming boats, 200–1; use of decked boats, 200–2, 204
Shetland haddock fishing, 204, 209, 217; season of fishing, 204 5; use of small boats, 205
Shetland herring fishing, 138, 153; adoption of decked boats, 201–2, 205–6, 207; curing, 139; curing stations, 203; earnings, 139, 204; engagements, 87, 201, 204, 206; fishing stations, 139; influx of east coast boats, 87, 153, 201; investment by mainland curers, 87, 202–3; markets, 139; scale of fishing, 204, 212; seasons of fishing, 87
Sixareen, 201; design, 129; use in herring fishing, 139, 201–2
Small-line fishing, 15–16, 176; see also haddock fishing
Smith's Knoll, 171
South Ronaldsay, 126–7
Spain, 60, 131
Spey, 9
Staxigoe, 29
Steam drifters, 150–1; bring new social relationships, 155; costs, 150, 213–14; earnings, 160–2; effect on farmer-fishermen, 182, 194, 207–8; ownership, 155, 157, 165; profits, 151, 159,

Steam drifters—*contd.*
213–14; share system, 157–8, 160; *see also* Aberdeen; Fraserburgh; Peterhead
Steamships, 183, 190–1, 193
Stettin, 55–6, 61–2
Stonehaven, 76, 85, 169, 176
Stornoway, 86, 115, 152, 183, 192–3
Stornoway fishery district, 195
Strome Ferry, 183, 186
Stromness, 124, 127
Stronsay, 126
Sutherland, 3, 36, 75, 103, 111, 114, 116, 191–2
Sutherland estates, 36
Sweden, 60–1

Tarbert, 197, 199
Tariffs, 60
Tay, 65, 163
Torry, 162, 165, 172
Transport costs, from Aberdeen, 172; from North West, 183, 186
Trawling, beginnings in Scotland, 167; companies, 179; composition of catch, 172–3; costs, 173–4, 216; earnings, 177–8, 216; fishing grounds, 167, 170–1, 173, 216; methods, 166–7; ownership and control, 178, 180, 216; recruitment of labour, 175–7; resistance by fishermen, 167; share system, 177–8; sources of capital, 179; terms of employment, 177; vessels, 173; wage-earning basis, 177; yields, 173–4, 216; *see also* Aberdeen

Ullapool, 6, 191
Union Harbour, Anstruther, 75, 91

Villages, composition and size, 10; geographical distribution, 9, 10, 12; growth of population, 48, 80; investment in steam drifters, 165; isolation of, 14, 80–1; livelihood, 163; partial decline, 162–3, 165, 175–6, 211–12, 217; use of family labour, 25; *see also* Buchan; Fife; Moray Firth

Walker, Thomas, 169, 178
Walls, 124
West Indies, 30, 50–2, 59, 139
Western Highlands and Islands Act, 1891, 182, 189
Whitby, 169
Wick, 29–30, 46, 158, 164–5, 200–1; concentration of herring fishing, 152; harbour development, 32–4, 74, 76–7; herring exports, 32, 51; immigration, 35–7, 81; large boats used in Shetland greatline fishing, 200; ownership of steam drifters, 157; scale of fishing, 39, 78, 152

Yarmouth, 153
Yorkshire, 87, 215

Zollverein, 60
Zulu class of boats, 82–3